ACS SYMPOSIUM SERIES **594**

Surfactant-Enhanced Subsurface Remediation

Emerging Technologies

David A. Sabatini, EDITOR
University of Oklahoma

Robert C. Knox, EDITOR
University of Oklahoma

Jeffrey H. Harwell, EDITOR
University of Oklahoma

Developed from a symposium sponsored
by the Division of Environmental Chemistry, Inc.,
and the Division of Colloid and Surface Chemistry
at the 207th National Meeting
of the American Chemical Society,
San Diego, California,
March 13–17, 1994

American Chemical Society, Washington, DC 1995

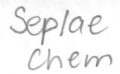

Seplae
Chem

Library of Congress Cataloging-in-Publication Data

Surfactant-enhanced subsurface remediation: emerging technologies /
David A. Sabatini, Robert C. Knox, Jeffrey H. Harwell, editors.

   p.   cm.—(ACS symposium series; 594)

Includes bibliographical references and indexes.

ISBN 0–8412–3225–3

   1. Soil remediation. 2. Groundwater—Purification. 3. Surface active
agents. I. Sabatini, David A., 1957– . II. Knox, Robert C.
III. Harwell, Jeffrey H., 1952– . IV. Series.

TD878.S85   1995
628.5'5—dc20   95–15245
                                                         CIP

This book is printed on acid-free, recycled paper.

Foreword

THE ACS SYMPOSIUM SERIES was first published in 1974 to provide a mechanism for publishing symposia quickly in book form. The purpose of this series is to publish comprehensive books developed from symposia, which are usually "snapshots in time" of the current research being done on a topic, plus some review material on the topic. For this reason, it is necessary that the papers be published as quickly as possible.

Before a symposium-based book is put under contract, the proposed table of contents is reviewed for appropriateness to the topic and for comprehensiveness of the collection. Some papers are excluded at this point, and others are added to round out the scope of the volume. In addition, a draft of each paper is peer-reviewed prior to final acceptance or rejection. This anonymous review process is supervised by the organizer(s) of the symposium, who become the editor(s) of the book. The authors then revise their papers according to the recommendations of both the reviewers and the editors, prepare camera-ready copy, and submit the final papers to the editors, who check that all necessary revisions have been made.

As a rule, only original research papers and original review papers are included in the volumes. Verbatim reproductions of previously published papers are not accepted.

M. Joan Comstock
Series Editor

Contents

vi

PANEL DISCUSSION

INDEXES

Preface

CHEMICAL RELEASES INTO THE SUBSURFACE are pervasive environmental problems. Sources of chemical releases range from abandoned hazardous waste disposal sites (e.g., Superfund sites) to leaking underground storage tanks at the corner gasoline station. Remediation of these releases has proven to be a formidable challenge. Many organizations are investing significant resources into the development of innovative technologies for expediting subsurface remediation.

Remediation efforts are frequently inhibited by an inability to extract contaminants from the subsurface due to the significant sorption of strongly hydrophobic chemicals (e.g., poly(chlorinated biphenyl)s) or due to the presence of separate phases of nonaqueous-phase liquids (NAPL; e.g., trichloroethylene). Surfactants may be used to enhance the solubility or to mobilize such contaminants. This approach has shown promise for significantly reducing the time and cost of remediation for sites contaminated with strongly hydrophobic or immiscible organic contaminants.

This volume combines a timely review of our current understanding of surfactant-based remediation technologies with an insightful discussion of critical issues that demand future consideration as these technologies move toward full-scale implementation. Contributors include representatives of academia, industry, and the regulatory community; each brings an important perspective that might be overlooked in a less comprehensive volume.

Recent research results are presented by leading experts in surfactant-based remediation technologies. The chapters in this volume range from fundamental discussions of physicochemical and biological processes affecting surfactant-based technologies to implementation and optimization issues affecting the widespread, full-scale utilization of these technologies. The last chapter summarizes a panel discussion on the general topic of the future of emerging surfactant-based remediation technologies. This book will thus be invaluable for scientists and engineers with research, teaching, consulting, and regulatory or management responsibilities for environmental remediation.

Acknowledgements

We thank Mary Walker of the ACS Division of Environmental Chemistry, Inc., for her professional and courteous assistance in organizing the symposium upon which this book is based and for her encouragement to produce this volume. We gratefully acknowledge financial support from the Division of Environmental Chemistry, Inc., for the two-day symposium in San Diego. We also acknowledge the Division of Colloid and Surface Chemistry for co-sponsoring the symposium.

We give special thanks to our invited speakers (Linda Abriola, Gary Pope, Richard Luthy, Abdul Abdul, and John Scamehorn) for their timely and informative presentations. We also acknowledge the presenters at the symposium and the authors of the chapters in this book for their significant contributions, their timely submittals, and their patience. We acknowledge the members of the discussion panel (James Greenshields, Linda Abriola, Abdul Abdul, Jeffrey Harwell and Candida West—moderator) for helping to make it a most profitable exercise. We thank the reviewers of the manuscripts in this volume, who were most cooperative and made valuable contributions to the quality of the chapters despite the short deadlines. We also thank the Schools of Civil Engineering and Environmental Science and Chemical Engineering and Materials Science, the College of Engineering, and the University of Oklahoma for their support of this activity.

Finally, we thank the ACS Books Department staff, and especially Rhonda Bitterli, for their assistance, patience, and encouragement in preparing this book and publishing it in a timely manner. We appreciate the professional approach and commitment to excellence of the ACS Books Department; it has once again been a pleasure working with them.

DAVID A. SABATINI
ROBERT C. KNOX
School of Civil Engineering and Environmental Science

JEFFREY H. HARWELL
School of Chemical Engineering and Materials Science

Institute for Applied Surfactant Research
University of Oklahoma
Norman, OK 73019

February 10, 1995

Chapter 1

Emerging Technologies in Surfactant-Enhanced Subsurface Remediation

David A. Sabatini[1,3], Robert C. Knox[1,3], and Jeffrey H. Harwell[2,3]

[1]School of Civil Engineering and Environmental Science,
[2]School of Chemical Engineering and Materials Science, and
[3]Institute for Applied Surfactant Research, University of Oklahoma,
Norman, OK 73019

Conventional pump-and-treat remediation has met with limited success, often due to the presence of residual saturation or strongly sorbed compounds. Innovative technologies are necessary to overcome these mass transfer limitations. Surfactant-based technologies have the potential to significantly enhance subsurface remediation. Examples of surfactant based technologies include enhanced contaminant extraction, improved bioavailability, and enhanced retardation via sorbing barriers. This chapter provides an introductory discussion of subsurface remediation, surfactant fundamentals, and surfactant-based remediation technologies. Finally, an overview of this volume (and the sessions from which it resulted) is provided.

Chemical releases into the subsurface environment are pervasive environmental problems; remediation of these scenarios has proven to be a formidable challenge. Cleanup of contaminated subsurface environments is complicated by the physical nature of the geologic formation, the behavior of chemicals introduced to the formation and heterogeneities inherent in these systems. Initial efforts to remediate contaminated ground water have met with mixed results, as discussed below.

Subsurface Remediation: Past Experience

Pump-and-treat remediation was initially prescribed for cleanup of subsurface contamination by both organic and inorganic contaminants. More recently, the limitations of this approach have been recognized. This is in part due to the realization that subsurface contamination exists in three zones: the source area (the original wastes or contaminated soil that continue to discharge into the ground water plume), the concentrated plume (center of mass of the ground water plume), and the dilute ground water plume. While the conventional pump-and-treat approach may effectively manage the dilute portion of the plume, innovative technologies are necessary to address the source zone materials (e.g., strongly sorbing organics -- PCBs, PAHs; trapped oil phases or residual saturation -- chlorinated solvents such as PCE, TCE). The inefficiency of conventional pump-and-treat methods has been highlighted

0097–6156/95/0594–0001$12.00/0

in several recent reviews. Keely (*1*) discusses factors limiting pump-and-treat remediation, including desorption of contaminants from media surfaces and dissolution of trapped immiscible phases. In a study of nineteen ongoing and completed ground water remediation systems, Haley et al. (*2*) delineated factors limiting subsurface remediation. While their analysis suggested that plume containment is commonly achieved with conventional pump-and-treat methods, they observed that mass recovery was typically much slower than originally anticipated. Again, desorption of highly hydrophobic compounds and dissolution of residual saturation are identified as limiting factors for conventional pump-and-treat approaches. Haley et al. (*2*) suggest that innovative methods be evaluated for overcoming these mass transfer limitations.

Introduction of chemical amendments for expediting pump-and-treat remediation of organic and inorganic compounds is discussed by Palmer and Fish (*3*). Chemical amendments enumerated include the following: complexing agents, cosolvents, surfactants via solubilization (micellar partitioning) and mobilization (ultra-low interfacial tension via middle phase microemulsion), oxidation-reduction agents, precipitation-dissolution reagents, and ionization reagents. An EPA-sponsored workshop considered technologies for expediting remediation of subsurface DNAPL contamination. Surfactant enhanced subsurface remediation was identified as a promising technology for expediting source-zone treatment (*4*).

Surfactant Fundamentals

The term surfactants comes from the descriptive phrase surface active agents. Surfactants are molecules that have both hydrophilic and lipophilic moieties. The amphiphilic nature of surfactant molecules causes them to accumulate at interfaces (e.g., air-water, oil-water, water-solid). For example, a surfactant will accumulate at an oil-water interface with its hydrophobic moiety (lipophilic tail) in the oil phase and its hydrophilic moiety (polar or ionic head) in the water phase; thus, both moieties of the molecule are in a preferred phase and the free energy of the system is minimized. Accumulation of surfactants at interfaces alters the nature of the interface, resulting in the designation of these molecules as surface active agents.

Surfactants are typically classified by the nature of their head group as cationic, anionic, nonionic and zwitterionic (both cationic and anionic groups). Surfactants are also characterized by their hydrophile-lipophile balance (HLB). Surfactants with a high HLB value are hydrophilic and thus more water soluble, whereas surfactants with low HLB values are lipophilic and thus more oil soluble.

A unique characteristic of surfactant molecules is their ability to self-assemble into dynamic aggregates known as micelles. This phenomenon occurs at elevated surfactant concentrations, upon satisfaction of interfaces and as the aqueous concentration (activity) increases. The surfactant concentration at which micelle formation commences is known as the critical micelle concentration (CMC). CMC values are a function of surfactant type (nonionics generally have lower CMCs than ionics), and system conditions (e.g., temperature, hardness) (*5*). Micelle formation distinguishes surfactants from amphiphilic molecules (e.g., alcohols) that exhibit a much lower degree of surface activity and do not form micelles.

Surfactant addition above the CMC will result in formation of additional micelles; i.e., the extramicellar surfactant concentration (the aqueous surfactant activity) is constant above the CMC. Micelles have a hydrophilic exterior (the hydrophilic heads are oriented to the exterior of the aggregate) and a hydrophobic interior (the hydrophobic tails are oriented towards the interior of the aggregate). Thus, micelles are analogous to dispersed oil drops; the hydrophobic interior of the micelle acts as an oil sink into which hydrophobic contaminants can partition. The increased "aqueous solubility" of organic compounds at supra-CMC surfactant concentrations is referred to as solubilization; as the surfactant concentration increases, additional micelles are formed and the contaminant solubility continues to increase.

Micelles are one of several possible surfactant phases. Surfactants with low HLB (that are oil soluble) will partition into the oil phase and may form reverse micelles. Reverse micelles have hydrophilic interiors and lipophilic exteriors; the resulting phenomenon is analogous to dispersed water drops in the oil phase. Surfactant systems intermediate between micelles (Winsor Type I system) and reverse micelles (Winsor Type II systems) can result in a third phase with properties (e.g., density) between oil and water. This third phase is referred to as a middle phase microemulsion (Winsor Type III system). The middle phase system is known to coincide with ultra-low interfacial tensions; thus, middle phase systems will greatly increase contaminant extraction from residual saturation. However, other mesophases may also occur in this transitionary region (e.g., liquid crystals--*5,6*).

The surface activity of surfactants causes them to accumulate at solid-liquid interfaces as well. In most cases surfactant sorption is undesirable; however, certain technologies are based on surfactant sorption (as discussed in the next section). Under typical aquifer conditions (near neutral pH), aquifer materials are generally negative in charge. As such, cationic surfactants are expected to be most susceptible to sorption. Surfactant sorption isotherms are typically favorable in nature (cooperative surfactant sorption results in three to four regions in the sorption isotherm; however, this phenomenon has been approximated by Langmuirian isotherms). The favorable nature of surfactant sorption is problematic when trying to elute surfactant from the subsurface but advantageous if surfactant sorption is desirable.

Surfactant precipitation is another loss mechanism of concern. Depending on the ionic composition of the subsurface system and the anionic surfactant utilized, precipitation may significantly impact the efficacy of the process. However, it is possible to select anionic surfactants that experience minimal precipitation. In addition, coinjection of nonionic surfactants will reduce the overall CMC, thereby reducing the surfactant precipitation realized (remember that the surfactant activity is a constant above the CMC). However, use of surfactant mixtures raises concerns as to potential chromatographic separation of the surfactants, which will potentially negate the desired effect of using a surfactant mixture.

Obviously, injection of surfactants into the subsurface will be allowed only when the resulting risk is deemed acceptable. Unfortunately, little is known about the migration and fate of surfactants in subsurface environments. Receptor exposure to surfactants may be deemed acceptable if these surfactants have direct food additive status (from the U.S. Food and Drug Administration). Also, exposure to surfactant metabolites may be deemed acceptable by virtue of the nature of these surfactants (many direct food additive surfactants combinations of fatty acids, sugars, etc.). Thus, by using direct food additive surfactants any risk associated with surfactant injection may be deemed acceptable, even under worst case scenarios for subsurface transport and fate processes. However, limiting surfactant selection to those having direct food additive status may be questionable, especially in light of the improved performance of other surfactants and the cost and nature of testing necessary for obtaining direct food additive status. Thus, other risk-based approaches for surfactant screening may be warranted as surfactant-based subsurface remediation technologies experience widespread implementation.

The intent of this section was to provide a brief introduction to surfactants and their properties. Many of these fundamentals are further developed and utilized in the chapters of this volume. Readers desiring an expanded discussion of surfactant fundamentals are directed to other references (*5-8*).

Surfactant Enhanced Subsurface Remediation

Having established that conventional pump-and-treat remediation is often limited by mass transfer constraints (desorption of strongly hydrophobic contaminants or dissolution of residual saturation), and having presented a basic introduction to surfactants and their properties, this section describes several innovative surfactant-based technologies for expediting subsurface remediation. The status of such emerging technologies is the subject of the remaining chapters in this volume.

A common goal of subsurface remediation is extraction of subsurface contaminants ("pump") with above ground treatment for waste processing and management ("and treat"). The hydrophobic sink provided by surfactant micelles can significantly increase the mass of contaminant extracted per volume of water pumped, thereby overcoming the mass transfer limitations historically experienced in conventional pump-and-treat efforts. As surfactant is added at concentrations exceeding the CMC, the number of micelles increases, thereby enhancing the contaminant solubility. This process (solubilization) will be equally appropriate for enhancing the desorption of highly hydrophobic contaminants and the dissolution of residual saturation. Mobilization (microemulsification) will have an even more dramatic effect in enhancing the extraction of residual saturation as the interfacial tension becomes negligible and the solubility enhancement escalates (the phases become virtually miscible). Thus, both these mechanisms can significantly enhance the efficiency of contaminant extraction over water alone.

An alternative to extracting contaminants from the subsurface is in situ management or in situ biodegradation of the contaminant plume. In situ management could include significant reductions in the rate of contaminant migration. Historically, hydraulic barriers were evaluated for this purpose (e.g., slurry walls, etc.); however, these systems were not the panacea they were originally envisioned to be. More recently the concept of sorbing barriers has been evaluated. Surfactant sorption can cause hydrophilic soil surfaces to become hydrophobic in nature (monolayer coverage--hemimicelles--renders the surface hydrophobic; bilayer coverage--admicelles--produces a hydrophilic exterior while creating a hydrophobic interior similar to the core of a micelle). The sorbed surfactant can significantly increase the sorptive capacity of the media for the contaminant (analogous to significantly increasing the fraction organic content), thereby dramatically reducing the mobility of the contaminant (effectively increasing the contaminant's retardation). While this approach greatly delays the appearance of the contaminant downgradient, it does not eliminate the potential of future risk. However, excavation of the sorbing barrier material or promotion of physicochemical and/or biological contaminant transformation by virtue of the increased detention time in the vicinity of the sorbing barrier will further minimize or eliminate future risk.

Solubilization has been perceived as a potential method for enhancing contaminant bioavailability via desorption or dissolution of residual / sorbed contaminants. At the same time, it is recognized that surfactant addition can have a multitude of other effects on the contaminant-microbial system which may either enhance or inhibit the rate and extent of biodegradation. Also of interest is the impact of surfactants on the degradation pathway (metabolites, etc.). Likewise, the biological fate of the surfactant itself is of interest; ideally the surfactant will be stabile enough to accomplish the task for which it is intended and yet labile enough that surfactant remaining in the subsurface is not a long term concern. Issues of cleanup standards and "how clean is clean enough" are common to all innovative remediation systems; while these techniques will greatly expedite mass removal/transformation, they are not a panacea for achieving drinking water standards throughout the source zone.

Subsurface heterogeneities pose a formidable challenge to all remediation efforts (conventional and innovative). These heterogeneities may occur in hydrodynamic, geochemical and/or contaminant properties of the system. A fundamental understanding of surfactant impacts on these processes (and vice versa) is paramount to successful implementation of these and other technologies.

Symposium Session / Symposium Series Book

The surfactant-based remediation technologies discussed above have significant potential for expediting subsurface remediation. While interest in these technologies is escalating (as evidenced by governmental and industrial research funding and an increasing number of feasibility studies being conducted by consulting firms), it is recognized that most of these technologies are still in the developmental stage (preliminary laboratory experiments to initial field trials). We thus perceived that the timing was right for convening experts from academia, industry, federal agencies, consulting firms and regulatory agencies to discuss the current state of surfactant-based technologies and to consider future directions for enhancing development and widespread utilization of these technologies. To this end, we organized a two day session at the 207th ACS National Meeting in San Diego, CA, March 13-17, 1994, entitled "Surfactant-Enhanced Remediation of Subsurface Contamination: Emerging Technologies."

This session consisted of twenty-six oral presentations and thirteen poster presentations which provided the developmental status of these technologies. Having thus established "where we are," the final activity was a panel discussion to assess "where do we go from here" with an emphasis on identifying critical factors for further technology development. The panel consisted of five members, two from academia, two from industry (one a technology user and one a surfactant manufacturer), and one from a regulatory agency (the USEPA).

Based on the valuable presentations and the insights generated during the panel discussion, and recognizing the need for a single manuscript to combine the work that would otherwise be published in a wide variety of forums (and disciplines), we proceeded with the challenge of developing the current volume. It was not possible to include all presentations since some authors were unable to participate and because ACS required us to minimize the number of chapters in this volume. Table I lists all the oral and poster presentations associated with the San Diego meeting. Noted in Table I are the five invited presentations; we would like to express our appreciation to Linda Abriola, Gary Pope, Richard Luthy, Abdul Abdul and John Scamehorn for their valuable presentations. We would also like to acknowledge the invited members of our panel discussion (James Greenshields, Linda Abriola, Abdul Abdul and Jeffrey Harwell) for their participation; special thanks go to the panel moderator, Candida Cook West, for agreeing to organize and moderate the panel (and for providing a summary of the panel discussion in this volume).

Following this introductory chapter, the volume has been divided into four sections: Enhanced Displacement Issues; Biotic / Biosurfactant Processes; Applications; and Panel Discussion. These divisions are provided in an attempt to improve the flow and organization of the volume. Obviously, these divisions are somewhat artificial; i.e., several chapters could easily be placed in more than one section. It is our hope that this volume serves as a useful compilation of the current status and future directions of surfactant-based environmental remediation technologies, and that it enhances the further development of these technologies.

Literature Cited

1.   Keeley, J. 1989. *Performance Evaluations of Pump and Treat Remediations.* USEPA, EPA/540/4-89-005. 19 pp.
2.   Haley, J. L., Hanson, B., Enfield, C., and Glass, J. *Ground Water Monitoring Review.* Winter 1991, 119-124.
3.   Palmer, C. D. and Fish, W. *Chemical Enhancements to Pump and Treat Remediation.* USEPA, EPA/540/S-92/001, 1992, 20 pp.

4.   USEPA (U.S. Environmental Protection Agency). 1992. *Dense Nonaqueous Phase Liquids -- A Workshop Summary*. USEPA, EPA/600/R-92/030.
5.   Rosen, M. J. 1989. *Surfactants and Interfacial Phenomena*. 2nd ed. John Wiley & Sons Inc., New York.
6.   Bourrel, M. and Schechter, R. S. *Microemulsions and Related Systems*. Surfactant Science Series, Vol. 30, Marcel Dekker, Inc., New York, 1988.
7.   Myers, D. 1992. *Surfactant Science and Technology*. 2nd ed. VCH Publishers, Inc., New York.
8.   West, C. C. and J. H. Harwell. 1992. *Environmental Science and Technology*. v. 26, no. 12, pp. 2324-2329.

Table I: Oral and Poster Presentations at the 207th ACS National Meeting in San Diego, CA, March 13-17, 1994

ORAL PRESENTATIONS
Impact of Surfactant Flushing on the Solubilization and Mobilization of Dense Nonaqueous Phase Liquids in Soil Columns. L. M. Abriola, K. D. Pennell and G. A. Pope. INVITED
Lessons from Enhanced Oil Recovery Research for Surfactant Enhanced Aquifer Remediation. G. A. Pope. INVITED
The Solubilization Properties of Nonionic Surfactants for In Situ Enhanced Removal of a Gasoline-Range Residual Hydrocarbon Phase from Contaminated Media. W. G. Rixey and R. A. Ettinger.
A QSAR for Solubilization of Nonpolar Compounds by Nonionic Surfactant Micelles. C. T. Jafvert and P. Van Hoof.
Surfactant-Enhanced Characterization of DNAPL Zones. G. W. Butler, R. E. Jackson and J. F. Pickens.
Enhanced Displacement of Tetrachloroethylene from Unsaturated Sediments by Surfactant Solutions. D. M. Tuck and P. R. Jaffe.
Impact of Surfactants on Subsurface Transport Properties of Immiscible Liquids. A. H. Demond, K. F. Hayes, D. Lord and A. Salehzadeh.
Sorption and Transport of a Nonionic Surfactant in a Sand-Aqueous System. Z. Adeel and R. G. Luthy. INVITED
Transport of Two Anionic Surfactants in an Unsaturated Loamy Soil. B. Allred and G. O. Brown.
Behavior of Linear Alkylbenzene Sulfonate in Oxygen Depleted Ground Water - Implications for In Situ Remediation. L. B. Barber, II, D. W. Metge, J. A. Field, and C. Krueger.

Surfactant Selection for Optimizing Surfactant Enhanced Subsurface Remediation. B. J. Shiau, J. D. Rouse, T. S. Soerens, D. A. Sabatini, and J. H. Harwell.

Sorption of Inorganic Oxyanions by Surfactant-Modified Zeolites. R. S. Bowman, E. J. Sullivan and G. M. Haggerty.

Use of a Model System to Explain the Inhibition of Naphthalene Biodegradation in the Presence of Several Nonionic Surfactants. J. R. Mihelcic, D. R. Lueking, M. I. Sottile and D. L. McNally.

Surfactant-Enhanced Mineralization of Polychlorinated Biphenyls and Polynuclear Aromatic Hydrocarbons. K. G. Robinson, M. M. Ghosh, W. Hunt, Z. Shi and I. Yeom.

Preliminary Studies on Surfactant Enhancement of Nitrate-Based Bioremediation. S. R. Hutchins, C. C. West and B. E. Wilson.

Biodegradation of Phenanthrene in the Presence of Nonionic Surfactants. S. Guha and P. R. Jaffe.

In-situ Surfactant Washing of Polychlorinated Biphenyls and Oils from a Contaminated Field Site: Phase II Pilot Study. A. S. Abdul and C. C. Ang. INVITED

Effectiveness of In-Situ Soil-Flushing with Surfactants - A Treatability Study. K. A. Bourbonais, G. C. Compeau and L. K. MacClellan.

Enhanced DNAPL Removal Using Surfactants: Capabilities and Limitations from Field Trials. J. C. Fountain, C. Waddell-Sheets, A. Lagowski, C. Taylor, D. Frazier and M. Byrne.

Modeling the Migration and Surfactant-Enhanced Remediation of PCE at the Borden Test Site Using the UTCHEM Compositional Simulator. G. A. Freeze, J. C. Fountain, G. A. Pope and R. E. Jackson.

Modeling the Effectiveness of Innovative Measures for Improving the Hydraulic Efficiency of Surfactant Injection and Recovery Systems. H. S. Gupta, Y. Chen and R. C. Knox.

Surfactant Recovery from Remediation of Subsurface Contamination. J. F. Scamehorn and S. D. Christian. INVITED

The Recycle/Reuse of Surfactant Solution Used to Remediate Contaminated Soil. A. N. Clarke, K. H. Oma, M. M. Megehee and D. J. Wilson.

Development of a Soil-Slurry Washing Technology Using Adsorbed Surfactant Aggregates. J. Park and P. R. Jaffe.

Direct Photoreduction of Hexachlorobenzene Solubilized within Surfactant Micelles. W. Chu and C. T. Jafvert.

Economic Considerations in Surfactant-Enhanced Pump-and-Treat. B. Krebbs-Yuill, J. H. Harwell, D. A. Sabatini and R. C. Knox.

POSTER PRESENTATIONS

Surfactant Enhanced Mobilization of M-Xylene in Soil Columns. J. W. Duggan, D. K. Ryan and C. J. Bruell.

Surfactant-Enhanced Solubilization of O-DCB, Dodecane, and PCE. A. M. Adinolfi, K. D. Pennell and L. M. Abriola.

Biosurfactant Enhanced Desorption of Cadmium from Soil. J. T. Champion, H. Tan, M. L. Brusseau, J. F. Artiola and R. M. Miller.

The Solubilization and Enhanced Transport of Low-Polarity Organic Compounds by Cyclodextrin. X. Wang, Q. Hu and M. L. Brusseau.

The Influence of Biosurfactant on Immiscible Liquid Displacement in Sand. G-Y. Bai, M. L. Brusseau and R. M. Miller.

Microemulsification of Single and Mixed Chlorinated Organic Solvents Using Edible Surfactants. B. J. Shiau, D. A. Sabatini, and J. H. Harwell.

Physical Modeling of Innovative Measures for Improving the Hydraulic Efficiency of Surfactant Injection and Recovery Systems. Y. Chen and R. C. Knox.

Experimental and Modeling Studies of Surfactant Enhanced DNAPL Remediation in 1-D Columns. T. S. Soerens, D. A. Sabatini and J. H. Harwell.

Influence of Surfactants on Microbial Degradation of Organic Compounds. J. D. Rouse, D. A. Sabatini, J. M. Suflita and J. H. Harwell.

Use of Twin-Head and Ethoxylated High Performance Anionic Surfactants in Subsurface Remediation. J. D. Rouse, D. A. Sabatini and J. H. Harwell.

Surfactant-Enhanced Desorption of TCE from Soil. J. J. Deitsch and J. A. Smith.

Modeling and Optimization of Ultrafiltration and Air Stripping for Use in Surfactant Recycling. K. M. Lipe, M. Hasegawa, D. A. Sabatini and J. H. Harwell.

Edible Hydrotropes in the Use of Edible Surfactants in Surfactant-Enhanced Soil Remediation. B. Krebs-Yuill, J. H. Harwell, D. A. Sabatini and R. C. Knox.

RECEIVED December 15, 1994

PHYSICOCHEMICAL PROCESSES

Chapter 2

Impact of Surfactant Flushing on the Solubilization and Mobilization of Dense Nonaqueous-Phase Liquids

L. M. Abriola[1], K. D. Pennell[1], G. A. Pope[2], T. J. Dekker[1], and D. J. Luning-Prak[1]

[1]Department of Civil and Environmental Engineering, University of Michigan, Ann Arbor, MI 48109
[2]Department of Petroleum Engineering, University of Texas, Austin, TX 78712

This paper provides an overview of on-going research related to surfactant-enhanced recovery of entrapped dense nonaqueous phase liquids (DNAPLs) in porous media. Three issues which may prove of particular importance to the design of successful field remediation schemes are highlighted: (1) rate-limited micellar solubilization; (2) the control of organic liquid mobilization; and (3) the influence of physical heterogeneities on organic distribution and recovery. Experimental and modeling investigations which explore each of these issues are presented and discussed. These studies reveal that (1) micellar solubilization is substantially rate-limited under flow conditions anticipated in engineered recovery schemes; (2) buoyancy forces may play an important role in DNAPL mobilization; and (3) entrapment of organic liquids in low permeability zones will strongly influence the performance of surfactant-enhanced solubilization operations at the field-scale.

Organic solvents and other petroleum-based products are frequently released to the environment as a separate organic phase or nonaqueous phase liquid (NAPL). When a NAPL migrates through the subsurface, capillary forces act to retain a portion of the organic liquid as discrete ganglia within the pores. These immobile ganglia may occupy between 5 and 40% of the pore volume at residual saturation (1, 2) and frequently represent a long-term source of aquifer contamination due to the low aqueous solubility of most NAPLs. Of particular concern are sites contaminated with dense NAPLs or DNAPLs. Such compounds, because of their large densities and low viscosities, are not typically confined to the unsaturated or capillary fringe zones. These dense liquids tend to migrate vertically under gravitational forces, and given sufficient spill volume, will displace water within the saturated zone and may spread deep within an aquifer formation.

It is now generally recognized that conventional pump-and-treat remediation methods are an ineffective and costly means for aquifer restoration when NAPLs are present (3, 4). The failure of this technique can be attributed, in large part, to the low aqueous solubilities of NAPLs and their relatively slow rates of dissolution. Over the past few years considerable interest has focused on surfactant flushing as an alternative method for recovering residual NAPLs from

0097–6156/95/0594–0010$12.00/0

contaminated aquifers (e.g., *5, 6*). This technique is based on the ability of surfactants to: (a) increase the aqueous solubility of NAPLs via micellar solubilization and (b) mobilize or displace the entrapped NAPL through reductions in the interfacial tension between the organic and aqueous phases. Surfactant washing has been successfully employed to remove sorbed or deposited polychlorinated biphenyls (PCBs) and polycyclic aromatic hydrocarbons (PAHs) from soil materials (*7, 8*). Initial applications of surfactant solutions to recover entrapped organic liquid contaminants, however, achieved mixed results (e.g., *9, 10*) . Recently, surfactant flushing has been successfully employed to remove automatic transmission fluid and residual dodecane from soil columns (*11, 12*). To date, relatively few studies have addressed the use of surfactants to recover DNAPLs from aquifer materials. Pennell *et al.* (*13*) reported that mixtures of sodium sulfosuccinate surfactants were capable of removing more than 99% of the residual tetrachloroethylene (PCE) from soil columns packed with F-95 Ottawa sand.

This paper presents an overview of our recent experimental and modeling research on surfactant enhanced remediation of DNAPLs. Some important issues which may have a significant impact on the field application of surfactant technologies to DNAPLs are highlighted. Specifically, these issues include: (a) rate-limited micellar solubilization; (b) the onset and extent of DNAPL mobilization during surfactant flushing; and (c) the influence of formation heterogeneity on DNAPL distribution and recovery. The presentation focuses on a single DNAPL, PCE, which was selected as representative of the chlorinated solvents typically encountered at contaminated sites. Illustrative data are presented relating to the application of several commercially-available surfactant formulations, including polyoxyethylene (POE) (20) sorbitan monooleate (Witconol 2722), sodium diamyl sulfosuccinate (Aerosol AY 100); sodium dioctyl sulfosuccinate (Aerosol OT 100), and sodium dihexyl sulfosuccinate (Aerosol MA 100). These surfactants were selected to produce a range of desired interfacial tensions between the surfactant solution and PCE, and for their phase behavior and capacity to solubilize PCE.

Micellar Solubilization

Although equilibrium batch solubilization measurements are useful for screening surfactants, it is important to recognize that such batch measurements may not be adequate for the prediction of surfactant performance in natural porous media. Previous soil column experiments in our laboratories have revealed that the micellar solubilization process is often rate-limited. We have observed substantial mass transfer rate-limitations for solubilization of residual dodecane and PCE in silica sands when flushing with aqueous solutions of Witconol 2722 (*12, 13*). The CMC for Witconol 2722 is reported to be 13 mg/L (*12*).

To further investigate solubilization rates, a series of batch and column experiments was conducted following the general procedures described by Pennell *et al.* (*13*). In the batch studies, an excess amount of PCE was contacted with a 4% (wt) solution of Witconol 2722 in 25-mL centrifuge tubes, which were gently mixed on a reciprocating shaker. The aqueous phase was destructively sampled over time and analyzed for PCE using a direct injection gas chromatography method developed for aqueous samples containing surfactant (*14*). The solubilized concentration of PCE was found to increase with time, approaching an equilibrium value of 38,500 mg/L after approximately 24 hours of mixing. The log K_m for this system was computed as 4.73 and the MSR as 7.18. The sharp rise and asymptotic plateau of the solubilization rate curve was similar to that reported by Arytyunyan and Beileryan for the solubilization of hydrocarbons in solutions of sodium pentadecylsulfonate (*15*).

Soil column experiments were then performed to explore the rate of PCE solubilization in a natural sandy porous medium. Borosilicate glass columns (5 cm i.d.) were packed with Oil Creek sand, supplied by the R.S. Kerr Environmental Research Laboratory, and saturated with de-aired water. Residual PCE saturations were established by injecting PCE liquid into water-saturated soil columns in an upflow mode, and then displacing the free product with water, in a downflow mode. Following the entrapment of PCE, a 4% solution of Witconol 2722 was pumped through the column and effluent samples were analyzed by the GC method described above.

The results of a representative surfactant flushing experiment are shown in Figure 1. At a Darcy velocity of ~3.4 cm/hr, the effluent concentration of PCE was found to approach a steady-state value of ~30,000 mg/L, which is ~8,500 mg/L less than the batch measured equilibrium value. To further investigate rate-limited solubilization of PCE, a flow interruption procedure was employed. Flow interruption data in Figure 1 suggest that equilibrium was attained within 18.5 hours. Note that subsequent interruptions of flow for 94, 20, and 69 hours yielded effluent concentrations in excess of the equilibrium value. This behavior was attributed to the formation of unstable macroemulsions which led to elevated concentrations of PCE in the aqueous phase. The existence of macroemulsions was indicated by visual examination of the effluent samples. Centrifugation of these samples at 7,500 rpm (23 cm dia. SORVALL SS34 rotor) was sufficient to break the macroemulsion, and the resulting aqueous phase concentrations were consistent with the equilibrium value (triangles shown in Figure 1). Figure 2 shows the results of an experiment conducted to assess the influence of flow velocity on the steady-state concentration of PCE exiting the column. These data indicate that the rate of residual PCE solubilization was insensitive to velocity over the range of Darcy velocities employed (1.6 to 8.3 cm/hr).

Results of these experiments may be compared with a similar set of experiments conducted for solubilization of dodecane in a quartz sand (13). For this system, the log K_m was computed as 9.02 and the MSR as 0.69. Effluent concentrations of dodecane were shown to approach equilibrium only after flow was interrupted for 100 hours. Concentrations exhibited some sensitivity to flow rate as the Darcy velocity was increased from 2 to 8 cm/hr (Figure 3). The greater sensitivity of dodecane concentrations to flow interruption and flow velocity indicate that dodecane was solubilized at a slower rate than PCE. These findings are consistent with data presented by Carroll and Ward, who found that the rate of solubilization increases with the polarity of the organic solute (16,17). Two mechanistic models, incorporating several coupled processes, have been proposed to describe rate-limited solubilization of organic liquids. The first model involves dissolution of the organic liquid into the aqueous phase and adsorption at the micelle-water interface, followed by diffusion into surfactant micelles (e.g., 15). In the second model, surfactant micelles are thought to diffuse to the organic-water interface, dissociate into monomers which are then adsorbed at the interface, and reform into micelles containing the organic liquids (e.g., 16,18). Further investigations are underway to characterize the solubilization process in more detail in order to ascertain the appropriateness of each modeling approach, to identify rate-limiting steps, to determine optimal surfactant concentration, and to develop predictive tools for flushing performance.

NAPL Mobilization

Although the discussion above focuses on micellar solubilization, aqueous surfactant solutions also have the capacity to displace or mobilize residual NAPLs from porous media. This process has been shown to be an efficient means for recovering residual PCE from soil columns (5, 13). For example, Pennell et al.

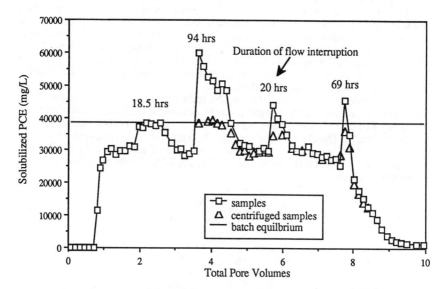

Figure 1. Effect of flow interruption on effluent concentrations of PCE during flushing of Oil Creek sand with a 4% solution of POE (20) sorbitan monooleate.

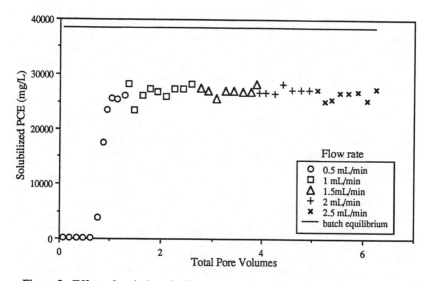

Figure 2. Effect of variations in flow rate on the effluent concentration of PCE during flushing of Oil Creek sand with a 4% solution of POE (20) sorbitan monooleate.

(13) reported that after injecting less than 2 pore volumes of a 4% Aerosol AY/OT solution, more than 99% of the residual PCE was removed from the column, with 80% being displaced as a separate organic phase. Despite the obvious potential of this recovery approach, the implications of this process on surfactant-based remediation scenarios must be carefully evaluated. Of particular concern is the possible downward migration of DNAPLs through an aquifer formation. Thus, it is essential to develop a means for evaluating the onset and extent of NAPL mobilization during surfactant flushing.

To induce NAPL mobilization, the reduction in interfacial tension (IFT) between the aqueous and organic phases must be sufficient to overcome the capillary forces acting to retain organic liquids within a porous medium. The capillary number (N_{Ca}) and Bond number (N_B) are dimensionless groups that can be employed to assess the impact of viscous and buoyancy forces on the mobilization of NAPLs in porous media *(13, 19, 20)*. These expressions can be defined as follows:

$$N_{Ca} = \frac{kk_{rw}\rho_w g\Delta\phi}{\sigma_{ow}\Delta x} = \frac{\mu_w q}{\sigma_{ow}} \tag{1}$$

$$N_B = \frac{\Delta\rho g k k_{rw}}{\sigma_{ow}} \tag{2}$$

where k is the intrinsic permeability; k_{rw} is the relative permeability to water; ρ_w is the density of water; g is the acceleration due to gravity; $\Delta\phi/\Delta x$ is the piezometric gradient; σ_{ow} is the interfacial tension between the organic and aqueous phases; μ_w is the viscosity of water; q is the Darcy velocity; and $\Delta\rho$ is the difference in fluid densities. When column experiments are conducted in a vertical orientation, with NAPL displacement in the direction of the buoyancy force, the Bond and capillary numbers may be superposed *(19)*. In order to sum these dimensionless groups, however, they must be written in a consistent form. Thus, the Bond number (eq. 1) was expressed in terms of the effective permeability, $k_e = k \, k_{rw}$, rather than the particle radius.

Buoyancy forces have been demonstrated to have negligible influence on surfactant enhanced recovery of petroleum in oil reservoirs and are generally neglected in predictive models of oil recovery *(21)*. Our experiments, however, indicate that buoyancy forces may play a major or even dominant role in DNAPL mobilization in aquifer systems. A series of soil column experiments was conducted to quantify the onset of PCE mobilization in Ottawa sand and to evaluate the utility of the capillary and Bond numbers. Residual saturations of PCE were established in soil columns following procedures similar to those described for the solubilization experiments. After the entrapment of PCE and initial water flushing, aqueous surfactant solutions (4% wt.) of Witconol 2722, Aerosol MA/OT 100 and Aerosol AY/OT 100 were pumped through a column in a downflow mode. This allowed for a sequential reduction in the IFT between PCE and the aqueous phase from 47.8 to 0.09 dyne/cm. To minimize the removal of mass due to micellar solubilization, only 1.5 pore volumes of each solution was injected through the column.

The PCE desaturation curve for 20-30 mesh Ottawa sand, expressed in terms of the capillary and Bond numbers, is shown in Figure 4. Here mobilization effects were estimated by measurement of the mass of displaced free product, which was large relative to the mass solubilized in the aqueous phase. Note that the PCE saturation remained essentially constant until the IFT was lowered to

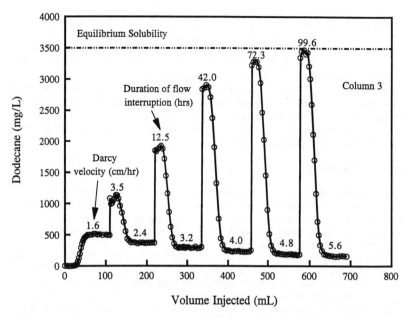

Figure 3. Impact of variations in flow velocity and duration of flow interruption on the recovery of dodecane from soil columns containing 20-30 mesh Ottawa sand (adapted from *(12)*).

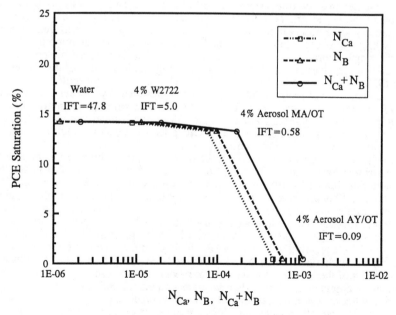

Figure 4. Tetrachloroethylene desaturation curve for 20-30 mesh Ottawa sand.

0.09 dyne/cm with the Aerosol AY/OT solution. During this phase of the experiment the Bond number was slightly greater than the capillary number, revealing the major importance of buoyancy forces in the displacement of PCE from the column. The sum of the capillary and Bond numbers required to induce PCE mobilization was ~2 X 10^{-4}, while almost complete removal was achieved at ~1 X 10^{-3}. Predictions based upon the capillary number alone would have been inadequate to characterize the mobilization process. It should be recognized that the critical sum required for mobilization is system specific and may vary by an order of magnitude depending upon the experimental design and the properties of the organic liquid and porous medium.

Numerical Modeling

Mathematical models can be developed and used to explore the potential impact of various physical and chemical processes on the performance of surfactant enhanced aquifer remediation (SEAR) at the field scale. Such mathematical approaches are necessarily limited by our understanding of processes and their interactions, and our ability to estimate appropriate model parameters. Although numerical simulators have been extensively employed in the petroleum literature to predict the performance of tertiary oil recovery schemes, mathematical models have only recently been applied to SEAR. Taking their lead from the petroleum literature, most of these modeling studies (22-25) have assumed local thermodynamic equilibrium among system constituents. As discussed above, however, there is a growing body of laboratory evidence to suggest that, at least for the micellar solubilization process, mass transfer rate limitations may be important.

   Abriola et al. (26) developed a conceptual model for surfactant enhanced solubilization, which incorporated mass transfer rate limitations. The model was implemented in a one-dimensional numerical simulator and was used to reproduce a series of surfactant column flushing experiments (12). Extrapolation of laboratory scale observations to the field, however, requires a multi-dimensional modeling approach. Multi-dimensional simulators can serve as important tools, permitting exploration of the potential influence of surfactant/organic properties, solubilization kinetics, aquifer formation heterogeneities, and flushing strategies on SEAR performance. The example model simulations presented below are based upon the sequential modeling approach presented by Dekker and Abriola (27). In this approach, the initial introduction and subsequent migration and entrapment of the organic pollutant in the subsurface is treated as an immiscible flow problem and modeled using VALOR, a two-dimensional multiphase flow simulator (28). Following entrapment and redistribution, dissolution and surfactant enhanced solubilization of the organic are then simulated using a two-dimensional extension of the model presented in Abriola et al. (26) under the assumption that no further free phase migration of the NAPL occurs. The contrasting time scales of the initial flow and entrapment process and the ensuing solubilization process make it reasonable to treat the two processes independently. A linear driving force expression is used to model mass transfer between the aqueous and organic phases. Estimates of NAPL-aqueous interfacial area are obtained by assuming spherical NAPL blob geometry, and interfacial area is decreased with time as the NAPL is solubilized. (See also 29-31 for further discussion of the interphase mass transfer model.) As described, the focus of this modeling approach is, thus, enhanced, rate-limited solubilization; simulation of surfactant-enhanced mobilization of the NAPL is precluded.

   The example model simulations presented below serve to illustrate model capabilities and highlight the influence of heterogeneity on SEAR effectiveness.

Consider the DNAPL spill event illustrated in Figure 5. Here a spill of PCE is simulated in a perfectly stratified saturated formation of fine sands. The simulation domain is composed of four layers of aquifer material with two contrasting permeability values k_1 and k_2, where $k_2 < k_1$. Porous medium properties are based on the well-characterized Borden aquifer (*32*). Entrapment and solubilization simulations were performed using two sets of permeability values. In the first, k_1 and k_2 were assigned values two standard deviations above and below the Borden site mean permeability value, respectively. In the second simulation, permeability values were taken from the extreme ends of the permeability distribution. In both simulations, the formation had a horizontal to vertical anisotropy ratio of 2:1. Capillary pressure parameters for the layers were modeled using a van Genuchten form of the capillary retention curve, scaled according to the layer permeabilities (*33*). Tables 1 and 2 summarize model input parameters.

For both simulations, PCE was introduced into the formation at a rate of 50 L/day for a period of 10 days, after which redistribution was allowed to occur under a remediation hydraulic gradient for a time period sufficient for the organic to become immobile. Figure 5 illustrates the pronounced impact of formation heterogeneity on the initial organic distribution. For the smaller permeability contrast, penetration into the lower permeability layer occurs after limited lateral spreading of the organic. Increasing contrast between layer permeabilities results in significant NAPL entrapment and spreading on top of the lower permeability layer, and very limited penetration into the lower strata.

Following migration and entrapment of the organic, a 0.02 m/m gradient was imposed on the domain, and a 4% solution of POE (20) sorbitan monooleate was introduced at the left side of the formation. An extraction well at the right side of the domain was used to withdraw the surfactant solution. Equilibrium micellar solubilities and organic/aqueous mass transfer coefficient expressions were derived from laboratory column studies (*12, 13*). The effect of formation heterogeneity on the solubilization process is illustrated in Figure 6, in which the NAPL remaining in the system is plotted versus the total flushing time.

It can be seen that the simulation employing a higher contrast in permeability has an initially higher removal rate, as the organic entrapped in the highly permeable top layer is readily accessible to the solubilizing surfactant solution and can be rapidly transferred into the aqueous phase. The small portion of NAPL located in the lower permeability layer is more resistant to solubilization; surfactant solution delivery to this layer is limited initially by diffusion from the more permeable layer and finally, by the slow advective transport rate in the low permeability layer. Consequently, the majority of the remediation time is expended in the removal of less than 1% of the total entrapped organic mass. This result can be compared with that of the simulation employing a lower permeability contrast. Here a large percentage of the total organic mass is entrapped in the lower permeability layer. In this case, however, under the same hydraulic gradient, the lower permeability layer is sufficiently permeable to allow the surfactant solution to contact the entrapped organic relatively rapidly and total clean-up time is reduced.

In perfectly stratified systems, such as those considered above, NAPL removal efficiency can also be shown to be highly sensitive to the assumed organic/aqueous mass transfer rate (*27*). Simulations reveal that, by varying the mass transfer coefficient over the range encountered in simple dissolution (*29,31*) and surfactant enhanced solubilization experiments, the estimated remediation time can be altered by a factor of two (*26*). These simulations also show that clean-up time is roughly proportional to the imposed gradient, while the total surfactant flushing volume required is relatively insensitive to this imposed gradient. It is also possible that surfactant sorption and desorption processes may

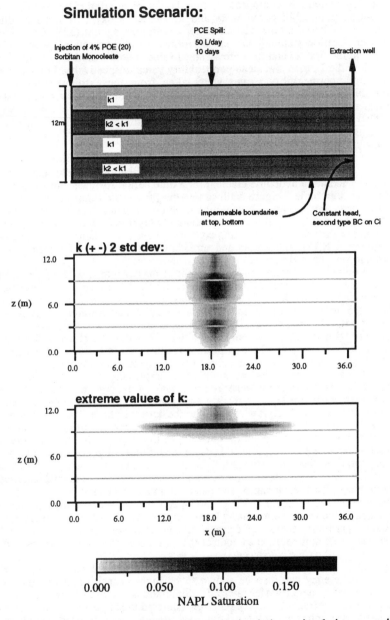

Figure 5. Two-dimensional PCE entrapment simulations: simulation scenario and results for two different permeability contrasts.

Table 1: Water and organic parameters used in numerical simulations.

Parameter	Symbol	Value	Units
Water density	ρ_w	999.0	kg/m^3
Water Viscosity	μ_w	0.112 x 10^{-2}	kg/ms
PCE Density	ρ_o	1625.0	kg/m^3
PCE Viscosity	μ_o	0.89 x 10^{-2}	kg/ms
Air/Water Interfacial Tension	γ_{aw}	72.0	dyne/cm
Air/organic Interfacial Tension	γ_{ao}	31.7	dyne/cm
Organic/Water Interfacial Tension	γ_{ow}	45.0	dyne/cm
PCE Diffusivity	D_{lo}	7.59 x 10^{-10}	m^2/sec
Residual Saturation PCE	$s_{res,o}$	0.17	
NAPL SA Factor	f	0.8	

Table 2: Soil matrix and permeability parameters used in numerical simulations.

Parameter	Symbol	Layers 1,3		Layers 2,4		Units
		k (+-) 2σ	extreme values of k	k (+-) 2σ	extreme values of k	
Porosity	n	0.34	0.34	0.34	0.34	
Horizontal Permeability	k_{xx}	0.56 x 10^{-10}	0.34 x 10^{-10}	0.52 x 10^{-11}	0.89 x 10^{-12}	m^2
Vertical Permeability	k_{yy}	0.28 x 10^{-10}	0.17 x 10^{-10}	0.26 x 10^{-11}	0.45 x 10^{-12}	m^2
Van Genuchten Parameters:	n	7.2	7.2	7.2	7.2	
	α	4.70 x 10^{-4}	1.30 x 10^{-1}	1.40 x 10^{-4}	2.1 x 10^{-2}	m s^2/kg
Longitudinal Dispersivity	α_L	0.3	0.3	0.3	0.3	m
Transverse Dispersivity	α_T	0.015	0.015	0.015	0.015	m

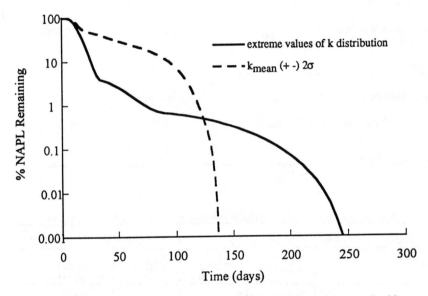

Figure 6. Effect of formation structure on PCE removal by surfactant flushing: NAPL removal curves for two different permeability contrasts.

have a significant impact on the overall performance of surfactant enhanced remediation, particularly in media of high organic carbon content. While sorption processes were not included in this preliminary work, future work will benefit from the added consideration of sorption effects.

Conclusions

The coupled experimental and modeling investigations described above reveal three important phenomena which may have substantial influence on surfactant remediation performance in the field. These phenomena are: rate-limited micellar solubilization, the onset and extent of NAPL mobilization, and the effects of physical heterogeneities. A thorough understanding of these phenomena will be required for effective remediation system design.

Laboratory studies have demonstrated that equilibrium micellar solubilities are not generally attained in soil flushing experiments employing solutions of POE (20) sorbitan monooleate. Interphase mass transfer rates are very weakly correlated to the system hydrodynamics, but do appear to depend strongly on organic contaminant hydrophobicity. The limited data suggest that such local mass transfer limitations may reduce the effectiveness of surfactant flushing in natural soils and that careful attention to fluid residence time will be required for system optimization. Further data will be needed to develop a mechanistic model of solubilization and to quantify mass transfer rates in a predictive sense. Solubilization experiments with chlorinated solvents have also demonstrated the potential for the formation of unstable macroemulsions. Although these macroemulsions were not observed to affect system hydrodynamics, the formation of such emulsions further complicates the development of predictive models.

Column experiments were conducted to investigate the onset of DNAPL mobilization. These investigations revealed that buoyancy forces can play an important role in the release of entrapped solvents, causing mobilization at higher interfacial tensions than would be expected based upon viscous forces alone. The use of the Bond and capillary numbers can provide a framework for the prediction of mobilization. Additional data will be required for the estimation of their critical sum as a function of porous media properties.

Mathematical models can be used to extrapolate the information gained from small-scale laboratory observations in homogeneous systems to more heterogeneous domains at the field scale. Such models can also help us to investigate the relationship between local interphase mass transfer rates and limitations to mass transfer created by variations in the physical properties of the medium. Preliminary simulations suggest that the effectiveness of surfactant-enhanced aquifer remediation will be limited by the volume of organic liquid entrapped in less accessible zones of lower permeability. For perfectly stratified systems, local interphase mass transfer rates may also have a substantial influence on NAPL recovery. Additional simulations are underway to explore the influence of more realistic heterogeneous permeability distributions on surfactant system performance.

Acknowledgments

Funding for this research was provided by the U.S. EPA Kerr Environmental Research Laboratory under Cooperative Agreement No. CA-818467, and the Great Lakes and Mid-Atlantic Hazardous Substance Research Center (HSRC) under Grant No. R-815750. Partial funding for HSRC research activities was also provided by the Michigan Department of Natural Resources and the U.S. Department of Energy. This paper has not been subject to Agency review and therefore does not necessarily reflect the views of the Agency, and no official endorsement should be inferred.

Literature Cited

(1) Schwille, F. In *Pollutants in Porous Media*; Ecology Studies; Springer-Verlag: New York, 1984; Vol. 47, pp 27-48.
(2) Wilson, J.L.; Conrad, S.H. In *Proceedings of NWWA Conference on Petroleum Hydrocarbons and Organic Chemicals in Ground Water*; National Well Water Association: Dublin, OH, 1984; pp 274-298.
(3) Mackay, D.M.; Cherry, J.A. *Environ. Sci. Technol.* **1989**, *23*, 630-636.
(4) Haley, J.L.; Hanson, B., Enfield, C; Glass, *J Groundwater Monitor.Rev.*, **1991**, *12*, 119-124.
(5) Fountain, J.C.; Klimek, A.; Beirkirch, M.G.; Middleton, T.M. *J. Harzard. Mat.* **1991**, 295-311.
(6) West,C; Harwell, J.H. *Environ. Sci. Technol.* **1992**, *26*, 2324-2330.
(7) Gannon, O.K; Bibring, P.; Raney, K.; Ward. A.; Wilson, J.; Underwood, J.L.; Debelak, K.A. *Sep. Sci. Technol.* **1989**, *24*, 1073-1094.
(8)Abdul, A.S.; Gibson, T.L. *Environ. Sci. Technol.* **1991**, *25*, 655-671.
(9)Nash, J.H. *Field Studies of In-Situ Washing*, U.S. Environmental Protection Agency: Cincinnati, OH, 1987; EPA/600/2-87/110.
(10) Zeigenfuss, P.S. M.Sc. Thesis, Rice University, 1987.
(11) Ang, C.C.; Abdul, A.S. *Ground Water Monit. Rev.* **1991**, 121-127.
(12) Pennell, K.D.; Abriola, L.M.; Weber, W.J., Jr. *Environ. Sci. Technol.* **1993**, *27*, 2332-2340.
(13) Pennell, K.D.; Jin, M.; Abriola, L.M.; Pope, G.A. *J Contamin. Hydrol.* **1994**, *16*, 35-53.
(14) Yavarski, T.P.; Pennell, K.D.; Abriola, L.M. In *Proceedings of the Pittsburgh Conference, Atlanta, GA,* 1993, p 271.
(15) Arytyunyan, R.S.; Beileryan, N.M. *Colloid J. USSR* **1983**, *45*, 355-360.
(16) Carroll, B.J. *J. Colloid Interface Sci.* **1981**, *79*, 126-135.
(17) Ward, A.J.I. *Proc. Royal Irish Acad.* **1989**, *89B*, 375-382.
(18) Chan, A.F.; Evans, D.F.; Cussler, E.L. *AIChE J.* **1976**, *22*, 1006-1012.
(19) Morrow, N.R.; Songkran, B. In *Surface Phenomenon in Enhanced Oil Recovery;* Shah, D.O., Ed.; Plenum Press: New York, 1981; pp 387-411.
(20) Ng, K.M.; Davis, H.T.; Scriven, L.E. *Chem. Eng. Sci.* **1978**, *33*, 1009-1017.
(21) Pope, G.A.; Baviere, M. In *Critical Reports on Applied Chemistry*, Baviere, P., Ed.; Elsevier Publishing: London, 1991, Vol 33.
(22) Wilson, D.J. *Sep. Sci.Technol.* **1989**, *24*, 863.
(23) Wilson, D.J.; Clark, A.N.*Sep. Sci.Technol.*, **1991**, *26*, 1177
(24) Harwell, J.H.; Sabatini, D.A.; Soerens, T.S. In *Migration and Fate of Pollutants in Soils and Subsoils, NATO ASI Series*, Series G, Petruzzelli, P.; Helfferich, F.G.; Eds.; Ecological Sciences, Springer-Verlag: Berlin, 1993, Vol. 32, pp 309-328.
(25) Brown, C.L.; Pope, G.A.; Abriola, L.M.; Sepehrnoori, K. *Water Resour. Res.*, in press, 1994.
(26) Abriola, L.M.; Dekker, T.J.; Pennell, K.D. *Environ. Sci. Technol.* **1993**, *27*, 2341-2355.
(27) Dekker, T.J.; Abriola, L.M. In *Toxic Substances and the Hydrologic Sciences,* Dutton, A.R., Ed.; American Institute of Hydrology, 1994, pp 290-301.
(28)Abriola, L.M.; Rathfelder, K.; Yadav, S; Maiza, M. *VALOR: A PC Code for Simulating Surface Immiscible Contaminant Transport,* Final Report #EPRI TR101018, Electric Power Research Institute, Palo Alto, CA, 1993.

(29) Powers, S.E.; Abriola, L.M.; Weber, W.J., Jr. *Water Resour. Res.* **1994**, *28*, 2691-2705.
(30) Powers, S.E., C.O. Loureiro, L.M. Abriola, and W.J. Weber, Jr., *Water Resour. Res.* **1991**, *27*, 463-477.
(31) Powers, S.E.; Abriola, L.M.; Dunkin, J. S.; Weber, W.J. Jr., *J. Contam. Hydrol.* **1994**, *16*, 1-33.
(32) Sudicky, E.A.*Water Resour. Res.* **1986**, *22*, 2069-2082.
(33) Leverett, M.C., *Trans. AIME*, **1941**, *132*, 149-171.

RECEIVED December 15, 1994

Chapter 3

A Quantitative Structure–Activity Relationship for Solubilization of Nonpolar Compounds by Nonionic Surfactant Micelles

Chad T. Jafvert[1], Wei Chu[1,3], and Patricia L. Van Hoof[2]

[1]School of Civil Engineering, Purdue University,
West Lafayette, IN 47907
[2]National Oceanic and Atmospheric Administration,
Great Lakes Environmental Laboratory, 2205 Commonwealth Avenue,
Ann Arbor, MI 48105

A simple semi-empirical equation is constructed that relates micelle-water partition coefficients (K_m) to octanol-water partition coefficients (K_{ow}) for nonpolar compounds solubilized in nonionic surfactant micelles. Combination of this K_m - K_{ow} quantitative structure activity relationship (QSAR) with other correlation, equilibrium, and mass balance equations allows for *a priori* estimation of solute distribution in surfactant-water-soil mixtures. These equations and experimental data on the distribution of hexachlorobenzene (HCB) and polychlorobiphenyls (PCBs) in micelle-water and micelle-water-soil solutions, respectively, are described. In addition, some initial experimental information is provided on surfactant-aided extraction of 2,3,4,5-tetrachlorobiphenyl from a sediment followed by photochemical reduction. For decontamination of soil containing aryl chlorides, this sequential process results in a both a recyclable surfactant solution and the destruction of the contaminants.

The distribution of nonpolar hydrophobic organic compounds (HOCs) between various surfactant micellar pseudophases, water, and soil has been examined and modeled. Distributions between the micellar pseudophases and water have been quantified for hexachlorobenzene (HCB) in solutions of eight different nonionic surfactants, with micelle-water partition coefficients, K_m (M^{-1}), calculated utilizing the chemical's solubility in distilled water. With this data and information previously published on other HOCs and surfactants *(1, 2)*, a simple semi-empirical equation is constructed that relates micelle-water partition coefficients, K_m, to octanol-water partition coefficients, K_{ow}. This equation takes the form: $K_m = \beta K_{ow}$, where the value of β is dependent

[3]Current Address: Department of Civil and Structural Engineering, Hong Kong Polytechnic, Hung Hom, Kowloon, Hong Kong

upon the specific surfactant. Combining the general form of this equation with other equations commonly invoked to describe the soil particle - water distribution of HOCs results in a predictive model useful for estimating the distribution of HOCs in surfactant-water-soil systems. In this chapter, we provide a brief review of these equations and compare calculated and experimental distributions. Finally, we present some information regarding additional factors that may affect surfactant recycling during proposed soil restoration schemes. More detailed descriptions of several aspects of this ongoing work are reported elsewhere *(ref. 3 and 4; Chu and Jafvert, Environ. Sci. & Technol., in press 1994; and Jafvert, et al., submitted for publication, 1994).*

Micelle-Water Partition Coefficients

Micelle-water partition coefficients, K_m, and octanol-water partition coefficients, K_{ow}, for HOCs are both inversely proportional to the concentration of chemical in the water phase. Hence, a relationship between K_m and K_{ow} may be defined. K_{ow} is defined by equation 1:

$$K_{ow} = \frac{C_o}{C_w^*} = \frac{\upsilon_o}{C_w^* \bar{V}_c} = \frac{\upsilon_o}{\upsilon_w^*} \tag{1}$$

where C_o and C_w^* are the concentrations of solute in the water-saturated octanol and octanol-saturated water phases, respectively (mol / L), υ_o is the volume fraction of chemical in the octanol phase, $\bar{V}_c$ is the molar volume of the solute, and υ_w^* is the volume fraction of chemical in the octanol-saturated water phase.

The K_m is defined according to equation 2:

$$K_m = \frac{C_m}{C_w S_m} = \frac{R}{C_w} = \frac{\upsilon_s \bar{V}_s}{C_w \bar{V}_c} = \frac{\upsilon_s \bar{V}_s}{\upsilon_w} \tag{2}$$

where C_m is the concentration of solute in the micelle phase based on total volume of solution, C_w is the distilled-water solubility, S_m is the concentration of surfactant in micellar form (M), R is the mole ratio of compound to surfactant in the micellar pseudo-phase, $\bar{V}_s$ and $\bar{V}_c$ are the molar volumes of surfactant and solute, respectively, and υ_s is the volume fraction of solute in micelles assuming dilute conditions. Note that C_w and υ_w in equation 2 are defined in terms of distilled water solubilities, whereas similar terms in equation 1 are defined in terms of octanol-saturated water.

Combining equations 1 and 2 results in equation 3:

$$K_m = \bar{V}_s \frac{\upsilon_s}{\upsilon_o} \frac{\upsilon_w^*}{\upsilon_w} K_{ow} = \beta K_{ow} \tag{3}$$

which is explicit for infinitely dilute systems. Some general observations can be made regarding this relationship. First, the ratio $\upsilon_w^* / \upsilon_w$ should remain constant and/or slightly increase with solute hydrophobicity. Second, for compounds which are solubilized in micelles and in octanol mainly through dispersive forces, the ratio υ_s / υ_o should be largely invariant with solute hydrophobicity. A slight decrease in this ratio may result from steric factors which may additionally attenuate the solubility of larger molecules in highly structured micelles of finite volume. Hence, the product of these two ratios may be nearly constant for a given surfactant, suggesting that a nearly constant K_m / K_{ow} ratio (defined as β) exists independent of the specific "hydrophobic" solute. This constancy was shown to be the case for values obtained for a series of PAH compounds with the anionic surfactant dodecylsulfate (3).

Third, K_m is proportional to surfactant molar volume, $\bar{V}_s$. The molar volume, however, is dependent upon the size of both the hydrophilic (e.g., polyethoxy group) and hydrophobic portion of a surfactant molecule. As the hydrophilic head group increases in size, the value of υ_s decreases, resulting in little change in K_m. Alternatively, as the hydrophobic portion increases in size the value of υ_s may be quite invariant, especially for longer chains, resulting in a direct linear correlation between K_m and $\bar{V}_s$. This latter effect is very similar to solubility in polymers in which an asymptotic limit to volume fraction solubility is rapidly approached as polymer length increases (5).

To account for these differences in solvation, we may subdivide $\bar{V}_s$ into the volumes of the specific surfactant functional groups and multiply each by the unique proportionality ($\approx (\upsilon_s \upsilon_w^*)/(\upsilon_o \upsilon_w)$) and the specific compound's K_{ow}, resulting in a series summation form of equation 3. For practical applications, we may formulate a less cumbersome equation by introducing surrogate parameters for functional group volumes. For example, carbon- or ethoxy- chain lengths may be used.

K_m - K_{ow} QSAR

Values of K_m (M^{-1}) were calculated from our data on HCB solubility in aqueous solutions containing either Tween 20, 80, or 85, Brij 30 or 35, polyoxyethylene-10-lauryl ether, Triton X-705, or Exxal F5715 (4), and from Kile and Chiou's data (1) on 1,2,3-trichlorobenzene (TCB) and p,p'-DDT solubility in aqueous solutions containing Triton X-100, X-114, or X-405, or Brij 35. The regression equations for our data are provided in Figure 1. The data for the regression curves are reported elsewhere (4), and have been omitted here for the purpose of clarity. Values of K_m were calculated from the slopes of these curves and the distilled water aqueous solubility of HCB.

To quantitatively compare the effect of surfactant functional groups on K_m, the coefficients (on L) contained in the following equation were determined utilizing the data set with a Levenburg-Marquardt algorithm (which minimized the sum of squared residuals between predicted and experimental values):

$$K_m\ (M^{-1}) = K_{ow}\ (\ 0.030\ L_1 - 0.0058\ L_2 - 0.0056\ L_3 + 0.0319\ L_4)\qquad(4)$$

where, L_1 = number of linear aliphatic carbons in the hydrophobic tail
L_2 = number of repeating ethoxy groups
L_3 = number of carbons in a sorbitan group (for the Tweens $L_3 = 6$)
L_4 = number of total carbons in alkylbenzene groups (for example, for the Triton series: $L_4 = 14$)

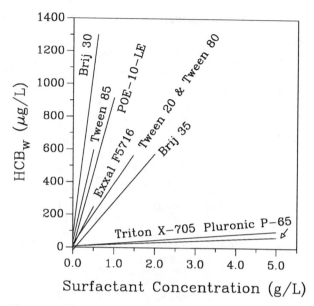

Figure 1. The Solubility of Hexachlorobenzene in Surfactant Solutions. The data for the regression curves are from reference *4*.

The similarity of some coefficient suggests that combining terms will result in a negligible increase in error:

$$K_m \ (M^{-1}) \ = \ K_{ow} \ (0.031 \ y_1 \ - \ 0.0058 \ y_2) \tag{5}$$

where, y_1 = number of "hydrophobic" carbons (aromatic or aliphatic, straight or branched reduced carbons)

y_2 = number of "hydrophilic" groups (sorbitan carbons or ethoxy groups).

We may now compare the experimental and calculated K_m values for the reported data. These data are provided in Table I *(1, 2, 6, 7)*. In general, the hydrophobicity of the solubilized chemicals is more significant than the structural differences among the surfactants examined in determining the magnitude of K_m for those surfactant-compound combinations examined. We also have calculated K_m values for pyrene, phenanthrene and naphthalene from the molar saturation data presented by Edwards *et al. (2)* for solutions containing either Brij 30, Triton X-100 or 2 additional nonionic surfactants and included this information in Table I. Because these other data were not used in determining the coefficients, they test the algorithm's predictive abilities.

Table I. Micelle-Water Partition Coefficients

Surfactant[1]	General Structure[2]	mol wt	Compound	log K_m (M^{-1}) exp.[3]	calc.	β	α/β[4]	α/β[5]
dodecylsulfate	$C_{12}SO_4^-$	265	HCB	4.33	-	0.068	1.0	1.0
Tween 20	$C_{12-18}S_6E_{20}$	1225	HCB	4.95	4.83	0.288	1.22	0.79
Tween 80	$C_{18}S_6E_{20}$	1310	HCB	5.00	5.10	0.320	1.12	1.0
Tween 85	$(3C_{18})S_6E_{20}$	1840	HCB	5.62	5.68	1.32	0.44	0.43
Brij 30	$C_{12}E_4$	362	HCB	5.18	5.04	0.472	0.24	0.20
			pyrene	4.85	4.71			
			phenanthrene	4.23	4.10			
			naphthalene	3.02	2.92			
POE 10-LE	$C_{12}E_{10}$	626	HCB	5.06	5.00	0.366	0.50	0.44
Brij 35	$C_{12}E_{23}$	1198	p,p'-DDT	5.83	5.72	0.223	1.5	1.37
			HCB	4.85	4.87			
			1,2,3-TCB	3.39	3.51			
Triton X-100	$B_8A_6E_{9.5}$	628	p,p'-DDT	5.95	5.94	0.379	0.50	0.42
			pyrene	4.55	4.75			
			phenanthrene	4.09	4.14			
			1,2,3-TCB	3.62	3.72			
			naphthalene	3.05	2.94			
Triton X-114	$B_8A_6E_{7.5}$	536	p,p'-DDT	5.91	5.95	0.391	0.43	0.35
			1,2,3-TCB	3.67	3.73			
Triton X-405	$B_8A_6E_{40}$	1966	p,p'-DDT	5.85	5.67	0.202	2.7	2.5
Triton X-705	$B_8A_6E_{70}$	3286	HCB	4.10	3.95	0.040	22	21
Exxal F5715	$B_{13}E_{10}$	640	HCB	4.80	5.04	0.204	0.93	0.80
Pluronic P-65	$P_{30}E_{40}$	3518	HCB	3.92	-			
Igepal CA-720	$C_8A_6E_{12}$	778	pyrene	4.63	4.73	0.364	0.60	0.55
			phenanthrene	4.06	4.12			
			naphthalene	3.03	2.92			
Tergitol NP-10	$C_9A_6E_{10.5}$	682	pyrene	4.76	4.78	0.404	0.51	0.43
			phenanthrene	4.25	4.17			
			naphthalene	3.09	2.97			

[1]Most names are trade names.

[2]Chemical formulas are those of the major component. C_n represents a saturated or unsaturated hydrocarbon chain of length n, B_n is a branched hydrocarbon chain, A_6 is a 6 carbon aromatic ring, E_n is n repeating $-CH_2CH_2O-$ groups, S_6 is a sorbitan ring, and P_n is repeating $-CH_2CH_3CH_2O-$ groups.

[3]Values for p,p'-DDT and 1,2,3-TCB were calculated from data presented by Kile and Chiou (1), with log K_{ow}'s of 6.36 and 4.14, respectively, from Chiou et al., (6); values for pyrene, phenanthrene, and naphthalene are calculated from molar saturation data presented by Edwards et al. (2) with log K_{ow} values of 5.17, 4.56, and 3.36 reported by Karickhoff, (7), and water solubilities of 1.0 x 10^{-6} M, 9.0 x 10^{-6} M, and 3.0 x 10^{-4} M, respectively, reported by Edwards et al. (2).

[4]This ratio is based on carbon-normalized masses: ((mass of solute / kg of soil organic carbon) / (mass of solute / kg surfactant carbon), hence, represents the ratio of the relative affinity of soil carbon to surfactant carbon for a solute.

[5]This ratio is based on total organic masses: ((mass of solute / kg of soil organic matter) / (mass of solute / kg surfactant).

Clearly, the utility of equation 5 as a predictive tool is evident as the experimental and calculatedvalues for the PAH compounds correspond to each other as accurately as those paired values used for the algorithm's construction. These additional data represent information on combinations of 3 additional compounds and 2 additional surfactants. In total, 13 surfactants and 6 compounds are represented in Table I, spanning four orders of magnitude in partition coefficients. The average absolute difference between measured and calculated log K_m values is 0.1 log units. This variation is remarkable given the simplicity of the model and the magnitude of errors generally associated with K_{ow} values alone.

Values for β for each surfactant are provided in Table I. Equations 3 and 5 illustrate that the value of β may be calculated either from the ratio K_m / K_{ow} (= β) or from the value 0.31 y_1 - 0.058 y_2 (= β). As will be described subsequently in more detail, a similar ratio defines nonpolar chemical distribution between soil organic carbon and water: α = K_{oc} / K_{ow}. Taking the ratio of these ratios (α / β) results in the proportionality K_{oc} / K_{ow}; or more specifically, the proportionality of solute concentration in the soil (or sediment) organic carbon to that in the micelle. This ratio can be made "dimensionless" ({μg / kg total organic matter} / {μg / kg surfactant}) by multiplying by 1,000 divided by the molecular weight of the surfactant, times a typically reported fraction of organic carbon within sediment organic matter (0.53). These ratios are reported in the last column of Table I for each surfactant. Alternatively, a similar ratio may be defined in terms of the carbon contents of the sediment and the surfactants, resulting in dimensionless units of the type: ({μg / kg sediment organic carbon} / {μg / kg surfactant carbon}). These ratios are reported in the next to last column of Table I for each surfactant. Typically, both ratios are close to 1.0, suggesting a similarity in the total solvation energies for solvation by soil organic matter (or carbon) and the surfactant micelles (or carbon). The major exception is Triton X-705 which is very ineffective on a mass basis due to the considerable proportion of the molecule that is nonpolar. The other values, however, imply that equal masses of soil organic matter and surfactant will lead to an approximately equal amount of a nonpolar chemical in each phase. Although order of magnitude deviations from this approximation exist, it should prove to be a useful "rule of thumb" for initial assessment of any surfactant micellar-based remediation process.

HOC Distribution in Saturated Soil Systems

Sorption to sediments or soils often correlates to the organic carbon content of a sedimentary material as the sorption process approaches equilibrium. Assuming a linear sorption isotherm, the organic carbon normalized partition coefficient, K_{oc} (L H_2O / kg organic carbon) simply is derived from the ratio of the solute concentration in the saturated soil or sediment organic carbon, $[C]_s/f_{oc}$ = $[C]_{oc}$ ({mol / kg soil}/{kg OC / kg soil}), to the solute concentration in the water phase, $[C]_w$:

$$K_{oc} = \frac{[C]_s}{[C]_w \ f_{oc}} = \frac{[C]_{oc}}{[C]_w} \tag{6}$$

Note that this relationship is limited by: *(i)* kinetic constraints, *(ii)* soil organic carbon and clay content, *(iii)* solute hydrophobicity and/or polarity, and *(iv)* solute concentration *(9)*.

For nonpolar solutes, K_{oc} values generally correlate to the respective solutes' K_{ow} values. Hassett *et al. (8)* found a near-zero intercept and a constant slope ($\alpha = 0.48$) for sorption of various PAHs onto 14 sediments or soils:

$$K_{oc} = \alpha K_{ow} \tag{7}$$

Upon amending soil slurries with surfactants, substantial sorption of surfactant molecules may increase the capacity of the soil to sorb other solutes. This process may be described as an equilibrium distribution of the solute between the aqueous phase and the mass of sorbed surfactant, $[S]_{ss}$ (mol / L):

$$K_s = \frac{[C]_{ss}}{[C]_w [S]_{ss}} \tag{8}$$

where $[C]_{ss}$ (mol / L of water) is solute associated with the sorbed surfactant.

Calculating f_{aq}. A useful and easily measured experimental parameter is the fraction of solute in the aqueous (water and micellar) phase, f_{aq}. This value may be expressed in terms of solute concentrations:

$$f_{aq} = \frac{[C]_w + [C]_{mic}}{([C]_{aq} + [C]_{mic} + ([C]_{oc} \cdot oc) + [C]_{ss})} \tag{9}$$

where oc (kg / L) is the soil organic carbon concentration, and the total mass balance on solute is represented by: $Mass_T = (V_{aq} ([C]_w + [C]_{mic} + [C]_{ss}) + M_s [C]_s)$. In this equation, V_{aq} is the total aqueous phase volume and M_s is the mass of soil material. Combination of equations 2 - 3 and 6 - 9 results in equation 10:

$$f_{aq} = \frac{1 + \beta K_{ow} [S]_{mic}}{(1 + \beta K_{ow} [S]_{mic} + \alpha K_{ow} oc + \beta K_{ow} [S]_{ss})} \tag{10}$$

For strongly sorbing solutes in systems with appreciable quantities of soil and surfactant, $[C]_{aq}$ is negligible; thus, equation 10 reduces to:

$$f_{aq} = \frac{[S]_{mic}}{([S]_{mic} + (\alpha / \beta) oc + [S]_{ss})} = \frac{[S]_{mic}}{([S]_{total} + (\alpha / \beta) oc)} \tag{11}$$

Note that equation 11 defines f_{aq} as independent of the specific solute. It should be recalled, however, that equation 11 was derived with several assumptions regarding the magnitude and mechanisms of distribution, notably that: (i) $[C]_w$ is insignificant from a mass balance standpoint, and (ii) solute distribution between water and soil, and water and micelles, is described by equations 2, 6, and 8. The independence of equation 11 from the specific solute results from the fact that compounds which obey these equations basically are partitioned between two "oily" or lipophilic phases (soil organic carbon and micelles) through nonspecific hydrophobic interactions. Compounds that meet these criteria include polycyclic aromatic hydrocarbons (PAHs) and polychlorinated biphenyls (PCBs) as well as many other environmentally relevant compounds.

Extraction from Sediments. This phenomenon (*i.e.*, independence from the specific solute) is clearly seen upon extraction of PCB-contaminated sediment with various surfactant solutions. We have performed such extractions with numerous surfactant solutions using a sediment sample provided by the U.S. EPA, Duluth, originating from near the mouth of the Ashtubula River in Ohio *(Jafvert et al., submitted for publication)*. The organic carbon content of this sediment was measured to be 1.5 %.

Sediment slurries were extracted with different surfactants for 24 hrs to determine the relative extraction efficiency of each surfactant. To quantify the total amount of each PCB congener in the sediment, several samples were prepared by combining 5 mL of the slurry (0.21 g dry sediment) with 5 mL acetone, sonicating for 30 min with a microprobe sonicator, centrifuging, and extracting 2 mL of the supernatant with 2 mL iso-octane containing 0.1 mg / L octachloronaphthalene as an internal standard. The results are shown in Figure 2 for samples extracted with Tween 80, with results reported as the fraction of each congener (*i.e.*, peak) recovered in the aqueous phase, f_{aq}. Evident in this figure is the nearly uniform recovery among the congeners (or peaks) for a given surfactant dose, indicating that distribution is largely independent of chlorine substitution on the biphenyl. Note that the standard deviation on relative recovery among the peaks is approximately 10 to 20 % of the experimental values, whereas for these same compounds, K_{ow} values span over 3 orders of magnitude. Recovery improves with increasing surfactant concentration, however, this increasing recovery diminishes at high surfactant dose. All of these phenomena are consistent with the model.

Equation 11 (the model) suggests that the recovery of hydrophobic organic compounds from the solid phase depends solely on the proportionalities α and β, the surfactant concentration in the micelles, $[S]_{mic}$, and the amount of soil organic carbon, *oc*. In our case, β (for each surfactant) and α (determined by Hassett *et al. (8)* to be 0.48) are constants, and $[S]_{mic}$ and *oc* are measured or measurable parameters. The value for β for Tween 80 is reported in Table I as 0.32. For these experiments, the amount of surfactant remaining in the aqueous phase was 0.28, 0.67, and 3.9 g / L for the 0.5, 1.0, and 4.0 g / L Tween 80 doses, respectively. With these values for α, β, f_{oc}, and $[S]_{mic}$, the recoveries of PCBs may be calculated with eq 11. For the 0.5, 1.0, and 4.0 g / L surfactant doses shown in Figure 2, the predicted (25, 41, and 84, respectively) and average experimental recoveries (13, 48, and 75, respectively) are in good agreement. An extensive compilation for other surfactants is provided elsewhere with quite similar results *(Jafvert et al., submitted for publication)*. Note that the

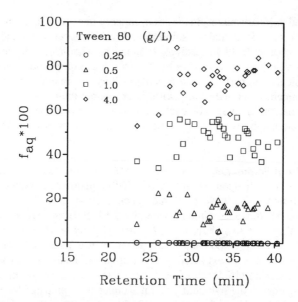

Figure 2. Recovery of PCB Peaks, After a 24 Hour Equilibration Period, as a Function of GC Retention Time and Tween 80 Dose from a 21 g / L Ashtabula River Sediment Slurry.

predicted recoveries where obtained utilizing only two easily measured system parameters ($[S]_{mic}$ and oc), and for cases where sorption of surfactant is negligible, $[S]_{mic}$ equals the amount of added surfactant. In some cases, $[S]_{mic}$, also, may be estimated with equilibrium phase distribution models.

We have not performed many experiments to define the kinetics of this process, however, in some basic experiments, sediment was extracted as a function of time to examine the rate at which the new equilibrium was reached. Samples were incubated for up to 30 days with no appreciable difference in recovery occurring from 12 hours (the shortest time examined) to 30 days. This equilibration time can be reduced, however, for the purpose of this work this was unnecessary.

The Resulting Hypothesis. To *estimate* the equilibrium distribution of any nonpolar organic compound in a soil-surfactant (micelle) system the only required surfactant-chemical dependent parameter is β. Experimentally determined values for β are listed in Table I for 14 surfactants. Values for additional nonionic surfactants may be estimated with equation 4 or 5. Hence, the equilibrium distribution of virtually any nonpolar organic compound in soil-water-surfactant systems may be estimated with these equations.

Practical Applications of Surfactant Extraction

Destruction of volatile and nonvolatile contaminants from groundwater, recovered through pump-and-treat operations, is routinely performed. The major impediment to these operations often identified is the slow release of the contaminants from the solid matrix, resulting in a voluminous amount of water over a long period of time that must be treated. To decrease both the volume of water and the time to site closure, several researchers have advanced the idea of adding surfactants to injection wells to facilitate the release of the contaminants. Clearly, bench scale studies such as those described above have shown surfactants to be effective in both decreasing the overall magnitude of adsorption and enhancing the rate of desorption. Thus, use of surfactants in above ground soil-washing operations potentially may enhance tremendously the performance of these processes. Surfactants are also very effective in decreasing the interfacial tension between water and oil phases. Hence, they have been used for some time as emulsifying agents to treat open-ocean oil spills. Work by several investigators is underway examining surfactant-aided recovery efficiencies of nonaqueous phase liquid (NAPL) spills; in such cases, recovery is improved due to this same effect on interfacial tension.

Although such treatment schemes have many promising features for *removal* of contaminants from complex matrices, a current obstacle to full scale implementation is the dearth of information regarding suitable treatment schemes for the resulting aqueous stream. Some notable suggestions for such schemes, however, are provided by other authors of this text. To be practical, treatment must result in both a purified water stream and recyclable surfactant. Four processes have been suggested as purification/recycle strategies. These include separation of the surfactant micelles and micellar solubilized contaminants from the aqueous stream by (low pressure) ultrafiltration, by gas stripping, or by solvent extraction, and alternatively, destruction of the contaminants within the surfactant solutions by photolytic processes *(Chu and Jafvert, accepted for publication in Environ. Sci. & Technol., 1994)*. These processes may be used alone on in combination.

Photolytic Processes. Use of photolytic processes on surfactant solutions, obviously is not limited to extracted contaminants. Several benefits may be achieved through the addition of low concentrations of surfactants to contaminated aqueous streams. In either case, performing photolytic reactions in surfactant micellar solutions has several advantages over homogeneous aqueous phase reactions. These include *(i)* the ability to attenuate undesirable side reactions, such as the formation of dimers from aromatic radical precursors, and *(ii)* the ability to increase the apparent reaction quantum yields. Regarding *(i)*, the distribution of contaminant molecules between water and the micellar pseudophase often results in a compartmentalization or isolation of micellized compounds from each other, such that dimerization reactions are infrequent. Solubilization by micelles also may reduce various quenching reactions by isolating micellized compounds from naturally occurring aqueous-phase quenching agents, such as oxygen and humic materials.

Regarding *(ii)*, several phenomena or reaction conditions lead to increases in the apparent quantum yields. The most obvious condition is the local environment within the surfactant micelle in which the contaminant molecule exists. A micelle is less polar than the aqueous phase, hence, reactions influenced by solvent polarity, will exhibit different rates in micellar systems compared to their rates in pure aqueous systems. For example, quantum yields for the photodechlorination of aryl chlorides are roughly an order of magnitude greater if the reactions are carried out in surfactant solutions or acetonitrile, as opposed to pure water, even with additional hydrogen sources present.

We have performed several photochemical experiments on surfactant-extracted soil contaminants with a RPR-100 Rayonet photochemical reactor equipped with a Merry-Go-Round apparatus and two 253.7 nm phosphor-coated low pressure mercury lamps (from Southern New England Ultraviolet). Contaminant functional groups shown to be reactive at this wavelength within micellar solutions include aryl halides and nitro aromatics. The aryl chlorides are reduced to lower chlorinated compounds or to other final products, depending upon the starting material. We have shown, for example, that hexachlorobenzene can be reduced totally to phenol and benzene, which decays further to unidentified final products *(Chu and Jafvert, accepted for publication in Environ. Sci. & Technol., 1994)*, and that photo-dechlorination of PCBs is possible under similar reaction conditions *(Chu and Jafvert, unpublished data)*. Similar results were obtained by Epling *et al. (10)*.

Combined Extraction and Photodecay. As stated previously, surfactant recycling is a critical issue. In theory, for many contaminants, soil extraction and photodecomposition performed in a serial manner has potential as a soil or sediment restoration technique. Results of a bench scale experiment are shown in Figure 3 in which an aqueous solution of Brij 58 was used to extract 2,3,4,5-tetrachlorobiphenyl (TeCB) from a spiked soil (EPA-11). This solution was subsequently centrifuged, irradiated at 254 nm, and used again as an extracting media for another fresh identically spiked sediment sample. This sequence of extraction (of new sediment each time) and irradiation was performed using the same surfactant solution four times. In this experiment, the initial 2,3,4,5-TeCB concentration was 1.4×10^{-6} mol / kg soil, resulting in an aqueous phase concentration of approximately 5.5×10^{-6} mol / L after each extraction. An irradiation time of 3 min was sufficient for over 99% photochemical reduction of the parent compound in a cell path of 1.3 cm. Although in this experiment we were interested in only the parent compound, in other studies, illumination for longer times resulted in complete dechlorination of the decay products. At the soil to surfactant ratio applied in this experiment (4.45), greater than 95 % of the TeCB was recovered from each new sediment sample in each of the first three extractions. Note that recovery decreases during the fourth extraction to around 80 %, and the optical density of the solution increases at the illuminating wavelength of 254 nm after each extraction. The latter phenomenon occurs due to the extraction of other substances by the micelles, presumably soil humic materials which absorb light at this wavelength. The loss in extraction efficiency during the fourth extraction results from the gradual loss of surfactant to the solid phase over the course of the serial extractions. Improved recovery and a stabilized optical density occurs if the solution is supplemented with a portion of new surfactant prior to each extraction. On the other hand, selection of an alternative surfactant that would tend to extract less soil material and tend to sorb less to the soil would similarly improve recovery of subsequent serial extractions. More

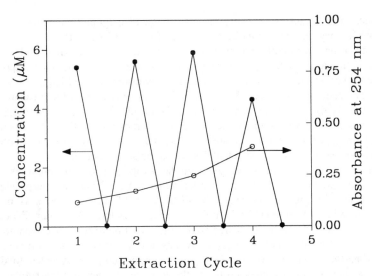

Figure 3. Extraction and Photolysis of 2,3,4,5-Tetrachlorobiphenyl (TeCB) from Soil with 100% Reuse of the Brij 58 Solution. Soil/surfactant ratio (w/w) was 4.45.

than likely, a trade-off between these two phenomena may occur. Extraction of large humic materials is likely enhanced by more polar functional groups on the surfactants through hydrogen bonding of these groups to the humics at the surface of the micelles. Alternatively, more polar surfactant functional groups may more effectively stabilize colloid size humic macromolecules in solution. Extraction of soil with dodecylsulfate rather than nonionic surfactants, for example, often results in more optically dense solutions. On the other hand, less polar (nonionic) surfactants will exhibit more sorption to the soil particles.

Clearly, an understanding of the predominant physical and chemical (and biological and toxicological) phenomena that are occurring in such complex systems will further facilitate evaluation and possible optimization of surfactant-aided remediation processes.

Summary

We have proposed a simple equilibrium model that should prove useful in evaluating many types of proposed surfactant-based soil remediation processes. The only parameters included in the model are those that represent the mass of each sorptive phase (oc, $[S]_{mic}$, and $[S]_{ss}$) and the solute-independent magnitudes of the distribution processes (α and β). This is possible because the "solvation" of nonpolar compounds by soil organic material and nonionic micelles occurs through the same hydrophobic mechanism, and in fact, is of nearly the same order of magnitude, resulting in nearly a constant recovery among nonpolar compounds for a given surfactant dose. With this model and a simple QSAR (equation 5) from which β may be calculated for most common nonionic surfactants, the distribution of virtually any nonpolar compound in a soil-water-surfactant solution may be estimated with only oc and $[S]_{mic}$ as measured values. The proportionality α / β (= K_{oc} / K_m) provides a quantitative measure of the solutes' affinity for soil organic carbon versus the micelles of a specific surfactant. As α is a constant and β varies among the surfactants, a range in relative affinities results. This ratio may be made dimensionless (mass solute / kg soil organic matter {or carbon}) / (mass solute / kg surfactant {or kg surfactant carbon}).

Most available nonionic surfactants are homologous series or mixtures. This includes the Tween, Brij, and Triton series surfactants. Hence, we purposely have avoided characterizing quantitatively the distribution of the surfactants as partition coefficients will vary among the homologs of each mixture. A similar difficulty is encountered when attempting to characterize solute sorption to sorbed surfactant, as the polarity of the sorbed surfactant mass likely depends on surfactant and soil concentrations. Clearly, different sorbed homolog distributions may be expected to display different tendencies in how they influence the sorption of other sorbing solutes.

Finally, some practical considerations of utilizing surfactant-aided soil extraction followed by photolytic destruction of the solubilized contaminants are suggested. Clearly, the most problematic phenomena to overcome, in both soil washing and aquifer flushing scenarios, is loss of surfactant to the solid matrix. In both scenarios, this phenomenon may prove to be a decisive factor, as the significance of this process may determine economic feasibility, and more importantly, future risks or liabilities as target contaminants are replaced by surfactant "contaminants". The widespread environmental distribution of surfactants (with the use of surfactants in household products and as emulsifying agents for agrichemicals as examples) suggests that risks may be small in comparison to benefits gained in some cases.

An additional benefit, not gained by many other existing strategies such as incineration, solidification, or isolation, is the possibility of site or soil "restoration" to some degree of its previous state. These other strategies "remediate" by controlling exposure, although, often result in property or a product of limited value.

Acknowledgements

The authors are indebted to Janice K. Heath, formerly of Technology Applications Inc., Athens, GA 30613, for her contribution in performing several of the laboratory experiments. A portion of this work was performed while C. Jafvert was a Research Environmental Engineer at the EPA Environmental Research Laboratory, Athens GA, and P. Van Hoof was a Post-Doctoral Research Associate at the University of Georgia. Partial support for P. Van Hoof was provided through the U.S. EPA Biosystems Program.

Literature Cited

1. Kile, D. E.; Chiou C. T. *Environ. Sci. Technol.* **1989**, 23, 832-838.
2. Edwards, D. A.; Luthy, R. G.; Liu, Z. *Environ. Sci. Technol.* **1991**, 25, 127-133.

3. Jafvert, C. T. *Environ. Sci. Technol.* **1991**, 25, 1039-1045.
4. Jafvert, C. T.; Heath, J. K.; Van Hoof, P. L. *Water Research,* **1994**, 28, 1009-1017.
5. Patterson, D.; Tewari, Y. B.; Schreiber, H. B.; Guillet, J. E. *Macromolecules,* **1971**, 4, 356-359.
6. Chiou, C. T.; Schmedding, D. W.; Manes, M. *Environ. Sci. Technol.* **1982**, 16, 4-10.
7. Karickhoff, S. *Chemosphere,* **1981**, 10, 833-846.
8. Hassett, J. J.; Means, J. C.; Barnwart, W. L.; and Wood, S. G. *Sorption properties of sediments and energy-related pollutants.* U.S. Environmental Protection Agency, Athens, GA., EPA-600/3-80-041, 1980.
9. Karickhoff, S. *J. Hydraulic Engineering,* **1984**, 110, 707-735.
10. Epling, G.A.; Florio, E.M.; Bourque, A.J.; Qian, X.H.; Stuart, J.D. *Environ. Sci. Technol.,* **1988**, 22, 952-956.

RECEIVED December 13, 1994

Chapter 4

Concentration-Dependent Regimes in Sorption and Transport of a Nonionic Surfactant in Sand–Aqueous Systems

Zafar Adeel and Richard G. Luthy

Department of Civil and Environmental Engineering,
Carnegie Mellon University, Pittsburgh, PA 15213

Sorption and transport of a nonionic surfactant in sand/aqueous systems appear to be controlled by concentration-dependent phenomena, resulting in two different sorption and transport regimes. Experiments were conducted in batch and column systems to evaluate the sorption isotherm and kinetics of sorption of Triton X-100 ($C_8PE_{9.5}$) onto Lincoln fine sand. The transition from one equilibrium sorption regime to the other occurred at an approximate surface coverage of 150 $Å^2$/molecule. An unusual two-step breakthrough curve was observed in column transport tests. An early surfactant breakthrough occurred at a fraction of the influent surfactant concentration; this was followed by a prolonged plateau in the effluent surfactant concentration. A transition from this plateau concentration to a second breakthrough segment was observed as surfactant surface coverage approached 180 $Å^2$/molecule. A two-stage empirical kinetic model for surfactant transport provided a reasonable fit to the experimental data.

The use of surfactants to assist removal of organic contaminants in *in-situ* aquifer remediation and *ex-situ* soil treatment systems has been discussed in recent papers (*1 – 3*). In this context, the principal points discussed in this chapter outline results from recent research, including the findings that: (i) nonionic surfactant sorption exhibits different characteristics for soil and aquifer sediment, (ii) the sorption of a hydrophobic organic compound (HOC) onto a solid is influenced by the sorption of nonionic surfactant, and appears to depend on the surface conformation of the sorbed surfactant, and (iii) the kinetics of nonionic surfactant transport through an aquifer sediment apparently are dependent also on sorbed surfactant conformation.

0097–6156/95/0594–0038$12.00/0
© 1995 American Chemical Society

The sorption of surfactant onto solid media is an important consideration in surfactant-assisted remediation for sediment-aqueous systems (*4, 5*). It is important to be able to understand and predict these sorption processes, as surfactant sorption onto sediment directly affects the partitioning of organic contaminants between the aqueous, micellar, and solid phases. Equilibrium sorption of nonionic surfactants onto silica, silica gels, soils and clean sands has been studied by various researchers, and the equilibrium sorption isotherms have been described by a variety of mathematical formulations (*4 - 10*). Recent research has indicated that the sorption phenomena controlling partitioning of a nonionic surfactant between natural media and aqueous phases are different for soils and sands. It is proposed that surfactant sorption may be governed by the amount of naturally-occurring organic matter associated with the solid phase, the mineral composition of the solid medium, and the surfactant concentration (*5, 11*).

Experimental evidence suggests that the conformation of nonionic surfactant molecules sorbed onto a surface may be important in affecting the degree of hydrophobicity of the surface, as well as affecting the affinity of the surface for sorbing HOC molecules (*5, 9, 10, 12*). In results summarized in this chapter, experiments comprising equilibrium sorption of Triton X-100 onto sand have shown that two distinct regimes of sorption may exist, and that such regimes may be dependent on the sorbed surfactant concentration. The two sorption regimes may correspond to difference in molecular conformations of the sorbed surfactant. In this regard, the extent of HOC sorption onto natural media containing sorbed nonionic surfactant at various concentrations has been used as an indicator of the degree of hydrophobicity of the surface. The data from HOC sorption experiments performed with various concentrations of a nonionic surfactant and an aquifer sand support the concept of different molecular conformations of sorbed surfactant and of two distinct sorption regimes.

Laboratory-scale column transport studies suggest that the conformation of sorbed surfactant molecules may play an important role in surfactant sorption kinetics. Abdul and Gibson (*13*) have reported that the degree of retardation for nonionic surfactant transport in laboratory-scale columns containing a natural sediment is dependent on the influent surfactant concentration. In the results reported in this chapter, experimental observations for transport of a nonionic surfactant in sand columns suggest that two different sorption regimes may control the transport process, in which surfactant sorption in each stage is governed by separate kinetic parameters. The two governing kinetic parameters appear to depend on the sorbed surfactant concentration and surfactant molecular conformation. This observation of two regimes is qualitatively supported by the corresponding equilibrium sorption data (*5*) and reported data for transport of rhodamine WT in alluvial sediments (*14*).

Experimental Procedures

Clean, Lincoln fine sand passing U.S. standard sieve no. 10 (2 mm) was used as a solid medium for batch sorption and column transport experiments. The properties of Lincoln fine sand are shown in Table I.

Table I. Properties of Lincoln fine sand.

Property	Units	Value
Organic carbon[a]	(g/g)	5×10^{-4}
Clay[b]	(g/g)	2×10^{-2}
pH (1:1 0.01M CaCl$_2$)[b]		6.4
CEC[b]		3.5
Bulk density[c]	(g/cm^3)	1.7
Surface area[d]	(m^2/g)	3.0

[a] Determined by Walkley-Black Method (15)
[b] From Wilson et al., (16)
[c] Determined for a packed column
[d] Measured according to BET method (17)

Triton X-100 ($C_8H_{17}-C_6H_4-O(CH_2CH_2O)_{9.5}H$) was selected as a representative nonionic surfactant due to its ability to enhance solubilization of organic compounds (4, 5), and because of this surfactant being studied by other researchers (6, 8 – 10, 18), and its availability in radio-labeled form. An aqueous solution containing ^{3}H-labeled and non-labeled Triton X-100 was used in column and batch experiments. Phenanthrene was obtained from Aldrich Chemical Company at 98% purity; ^{14}C-labeled phenanthrene was acquired from Amersham Corp. An aqueous solution containing ^{14}C-labeled and non-labeled phenanthrene was used in some batch sorption experiments. 0.01 M CaCl$_2$ was added to facilitate solid-liquid separation. The properties of Triton X-100 and phenanthrene are shown in Table II.

Table II. Properties of Triton X-100 and phenanthrene.

Property	Triton X-100	Phenanthrene
MW[a]	625	178
Solubility[a]	-	7×10^{-6} mol/L
log K_{ow}[a]	-	4.57
CMC[b]	1.8×10^{-4} mol/L	-
Aggregation No.[c]	140	-
Cloud Point[d]	65°C	-

[a] From Edwards et al., (5)
[b] From surface tension measurements
[c] From Robson and Dennis (18)
[d] From Partyka et al., (6)

Aqueous solutions were counted for ^{3}H or ^{14}C activity with 10 mL of Packard Optifluor scintillation cocktail and a Beckman LS 5000 TD liquid scintillation counter (LSC). The radioactivity was measured as disintigerations per minute per mL of liquid sample.

Batch surfactant sorption tests were conducted in 50 mL centrifuge tubes with PTFE septa, each containing 5 g of Lincoln fine sand and 30 mL of aqueous surfactant solution at various concentrations. For experiments with phenanthrene, an aqueous solution containing phenanthrene was equilibrated with the sand for 24 hours and the surfactant solution was then added to the tubes. The tubes were rotated end over end for at least 24 hours, followed by centrifuging at 1600 g for 30 min. In order to remove any suspended particles, the supernatant was expressed through Acrodisc PTFE filters (1 μm pore size), which were conditioned by wasting an initial 5 mL, and then directly dispensing the filtrate into LSC vials containing scintillation cocktail. These vials were counted for ^{3}H or ^{14}C activity and the sorbed concentrations of Triton X-100 or phenanthrene, respectively, were determined by mass balance.

One-dimensional column transport experiments were performed with a 7.53 cm long, 2.20 cm I.D., stainless steel column, packed with Lincoln fine sand in 18-20 layers. A number of pore volumes of de-ionized water containing 0.01 M $CaCl_2$ were flushed through column to condition the soil prior to pumping the surfactant solution. The aqueous surfactant solution was pumped from a stainless steel reservoir by an HPLC pump. The effluent surfactant solution from the column was accumulated in 10 mL tubes by an Eldex fraction collector; the volume of each sample collection depended on the flow rate and the duration of sampling. The ^{3}H activity in these samples was determined by LSC counting. Pumping of surfactant solution was discontinued after the effluent concentration became equal to the influent concentration. Clean, de-ionized water containing 0.01 M $CaCl_2$ was then pumped and surfactant desorption was measured in the manner described above for column effluent analysis.

Sorption of Surfactant and Phenanthrene

This section describes equilibrium sorption of Triton X-100 onto Lincoln fine sand as well as sorption of phenanthrene onto Lincoln fine sand in the presence of Triton X-100. A mathematical formulation describing equilibrium sorption of Triton X-100 is also presented.

Sorption of Nonionic Surfactant onto Sand.   In results described below, it is shown that Triton X-100 molecules may sorb directly onto the solid surface, or may interact with sorbed surfactant molecules, the sorption mechanism apparently being dependent on the nature of the sorbent and the surfactant dose (5). In the case of a mineral surface, or low-organic content aquifer sediment with very few sorbed surfactant molecules, the sorption of Triton X-100 surfactant molecules may occur mainly due to van der Waals interactions between the hydrophobic and the hydrophilic moieties of the surfactant and the surface (19). In comparison, at higher surfactant doses such sorption may occur through more structured surfactant arrangements including the formation of monomer surfactant clusters on the surface or the formation of admicelles or bilayers. These arrangements may be governed mainly by interactions between hydrophobic moieties of the surfactant molecules (5, 19). The surface arrangements of surfactant molecules may be patchy rather than uniform.

Figure 1 shows the sorption isotherm obtained from batch experiments for Triton X-100 and Lincoln fine sand. The data are expressed as the logarithm of sorbed phase surfactant concentration (mol/g solid) versus the logarithm of the bulk aqueous-phase surfactant concentration (mol/L). It is observed that sorption continues well beyond the point at which CMC occurs in the aqueous phase.

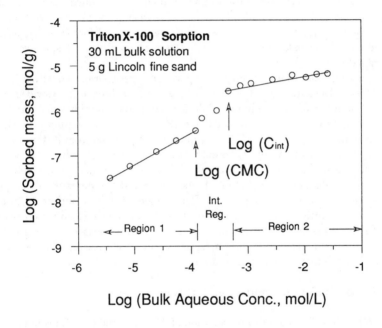

Figure 1. Batch experimental data for sorption of Triton X-100 onto Lincoln fine sand. (o) show experimental data.

The sorption data reveal some complex relationships. At low surfactant concentrations in region 1 the sorption is Freundlich-type, up to the point at which the critical micelle concentration (CMC) is reached in the aqueous phase. There is an intermediate region, beyond which the sorption appears to be Freundlich-type in region 2. The transition from the intermediate region to region 2 occurs at a bulk aqueous-phase surfactant concentration identified as C_{int}. The sorbed-phase surface concentration corresponding to C_{int} is computed to be approximately 150 Å^2/molecule. It is envisioned that the orientation of the sorbed surfactant molecules undergoes a transition from a more-or-less patchy flat-lying conformation to a patchy bilayer conformation as surfactant concentration is increased from region 1 to region 2 (*5, 11*). At the highest end of the sorption isotherm, i.e., that corresponding to an aqueous concentration of 150 times the CMC, the surface coverage is about 77 Å^2/molecule. This conceptualization of the orientation of sorbed surfactant molecules is illustrated schematically in Figure 2, where stages 1 and 2 correspond to regions 1 and 2 of Figure 1.

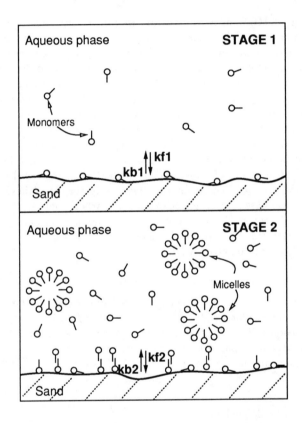

Figure 2. Schematic representation of the proposed model for sorption of Triton X-100 onto Lincoln fine sand.

Liu et al. (*4*) have studied sorption of Triton X-100 onto a soil; the results can be qualitatively compared to sorption of Triton X-100 onto Lincoln fine sand. Batch experiments by Liu et al. (*4*) for sorption of Triton X-100 onto Morton soil, having an organic carbon content of 0.96%, show that there is no significant sorption at aqueous surfactant concentrations greater than that at which the CMC occurs in the bulk aqueous phase. The surfactant sorption onto soil up to this dose can be described by a Freundlich-type linear isotherm (*4*), i.e., the plotted data for the logarithm of sorbed surfactant mass versus the logarithm of aqueous surfactant concentration can be fitted by a straight line. The differences in mineral composition and organic carbon fraction between Lincoln fine sand and Morton soil may account for the probable differences in the sorbed-phase conformations of the surfactant molecules. Surfactant sorption onto sand may occur through both solid-phase and sorbed-surfactant interactions, while sorption onto soil may occur through non-specific, hydrophobic interactions with soil humic matter.

Partitioning of HOC and Surfactant in Solid/Aqueous Systems. The data for partitioning of phenanthrene in the presence of a nonionic surfactant is presented here to corroborate the existence of different conformations of the sorbed surfactant molecules. It is proposed that the degree of hydrophobicity of the solid surface depends on the sorbed surfactant concentration.

Means et al., (20) have shown the sorption of polynuclear aromatic hydrocarbons (PAHs) onto 14 different materials is significantly correlated ($r^2 > 0.95$) with the organic carbon content of the solid matrix; the partition coefficients were found to be independent of all other solid properties. Sorbed surfactant may be considered to be an additional source of organic carbon. As a result, the capacity of a solid, such as Lincoln fine sand, to sorb HOC is increased by the sorbed surfactant. In order to account for this increase in sorption, the HOC sorption computations are based on an effective organic carbon content f_{oc}^*, which is the sum of naturally-occurring organic carbon and organic carbon associated with the sorbed surfactant. The mass of organic carbon associated with the sorbed surfactant, M_{surf}, can be given as:

$$M_{surf} = S_{surf} MW_{surf} f_c \tag{1}$$

where S_{surf} is the mass of sorbed surfactant per unit weight of the solid (mol/g), MW_{surf} is the molecular weight of the surfactant (g/mol), and f_c is the weight fraction of carbon in the surfactant. The effectiveness, ε, of sorbed surfactant as a sorbent for HOC compared to natural organic carbon is used to compute the effective organic carbon fraction, f_{oc}^*:

$$f_{oc}^* = f_{oc} + \varepsilon M_{surf} \tag{2}$$

where ε can be defined as:

$$\varepsilon = f_{oc} \frac{[K_{d,cmc} S_{cmc} - K_d S_w]}{K_d S_w M_{surf}} \tag{3}$$

The terms in equation 3 are described as follows. In general, the partitioning of HOCs between solid and aqueous phases in the presence of surfactant micelles and sorbed surfactant molecules may be described by two partition coefficients: K_m, for partitioning between micellar and aqueous pseudophases; and $K_{d,cmc}$, for partitioning of HOC between the solid and the aqueous phase in the presence of micelles.

$$K_m = \frac{X_{mic}}{X_{aq}} \tag{4}$$

$$K_{d,cmc} = K_d \left(\frac{S_w f_{oc}^*}{S_{cmc} f_{oc}} \right) \tag{5}$$

where X_{mic} and X_{aq} denote the mole fractions of HOC in the micellar and the aqueous pseudophases (mol/mol), S_w is the aqueous solubility of HOC (mol/L), S_{cmc} is the bulk aqueous solubility of the HOC in surfactant solution at the CMC (mol/L), f_{oc} is the organic carbon fraction naturally associated with the sediment (g/g), f_{oc}^* is the effective value of f_{oc} after surfactant sorption (g/g), and K_d is the HOC solid/aqueous partition coefficient (L of aq. solution / g of solid) in the absence of surfactant. Liu et al. (*4*) have shown for phenanthrene sorption onto Morton soil in the presence of micellar Triton X-100 the value of $K_{d,cmc}$ is constant at supra-CMC surfactant doses. The partitioning concepts described by equations 4 and 5 are schematically depicted in Figure 3 for the case of phenanthrene sorption onto sand in the presence of surfactant.

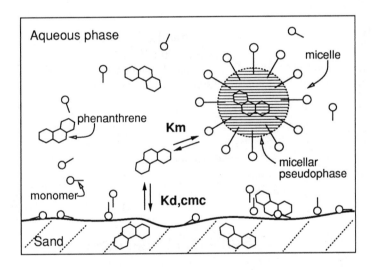

Figure 3. A schematic illustration of the partitioning of phenanthrene between the bulk aqueous and the solid sand phase in the presence of a micellar nonionic surfactant.

A value of unity for ε was found to be adequate for describing sorption of phenanthrene onto Morton soil in the presence of Triton X-100 (*21*). However, the value of ε for sorption of phenanthrene onto Lincoln fine sand in the presence of Triton X-100 varied depending on the sorbed surfactant concentration. This value was found to be always greater than unity, i.e., more than that for naturally-occurring organic matter on a carbon normalized basis, as shown in Figure 4 (*5*). It is surmised that this variability in HOC sorption is governed by the molecular conformation of the sorbed surfactant. At low concentrations, where surfactant coverage on the surface is sparse, the sorption of HOC onto Lincoln fine sand is enhanced to a relatively smaller degree. At higher surfactant concentrations approaching C_{int}, the enhancement in HOC sorption is 25 times that in the absence of any sorbed surfactant. This relative increase in HOC sorption may be attributable to the creation of a more hydrophobic interface between the sorbed surfactant molecules and the aqueous phase due to the alignment of surfactant hydrophobic moieties towards the aqueous phase. At still higher concentrations of the surfactant, the enhancement in HOC sorption is reduced to about 3–4 times that in the absence of sorbed surfactant. At such surfactant concentrations, the sorbed surfactant molecules may form clusters or bilayers with their hydrophilic moieties aligned towards the aqueous phase.

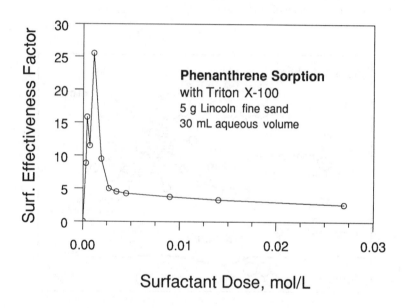

Figure 4. Computed values of the effectiveness of sorbed Triton X-100 surfactant for enhancing HOC sorption onto Lincoln fine sand.

Mathematical Formulation.   A unified mathematical expression describing the equilibrium surfactant sorption isotherm, rather than the discrete region-by-region Freundlich-type isotherms shown in Figure 1, may be useful in modeling the surfactant transport in sediment-aqueous system. Such a mathematical model, proposed by Gu and Zhu (8) for sorption of Triton X-100 onto a silica gel, can be used to describe the data presented in Figure 1. Consider a case where n monomers of the nonionic surfactant associate with the surface to form a hemi-micelle.

$$site + n\ monomer \rightleftharpoons hemi\text{-}micelle \tag{6}$$

The above reaction can be described mathematically by using the law of mass action giving the following isotherm expression (*8*):

$$S_s = g(C_s) = \frac{S_\infty K(C_s)^n}{[1 + K(C_s)^n]} \tag{7}$$

where S_s is the sorbed concentration of the surfactant on the surface (mol/g), S_∞ is the maximum sorbed concentration (mol/g), C_s is the aqueous-phase surfactant concentration (mol/L), and K is a constant (L/mol). Equation 7 may be envisioned as a combination of Langmuir and Freundlich isotherms. It describes limited sorption at the high end, like a Langmuir isotherm, and reduces to a Freundlich-type isotherm at low concentrations. The equilibrium sorption isotherm plotted according to equation 7 is shown in Figure 5 for sorbed concentrations (S_s) versus the aqueous concentrations (C_s).

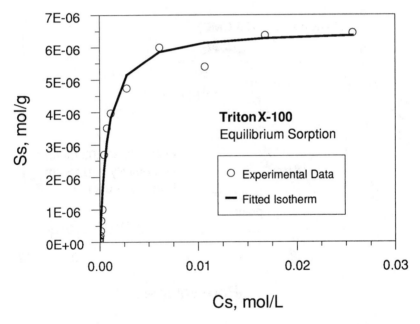

Figure 5. Batch experimental data and the fitted non-linear sorption model for Triton X-100 sorption onto Lincoln fine sand.

Surfactant Transport in Sand

This section presents the experimental data for transport of Triton X-100 in Lincoln fine sand. A sorption-kinetic modeling concept is proposed and its mathematic formulation is used to fit the experimental data. The transport of Triton X-100 in laboratory sand column tests is believed to be governed by two different sorption regimes and this behavior may be explained with the help of the equilibrium sorption data.

Experimental Observations. Several different column tests were conducted to evaluate surfactant transport at various influent surfactant concentrations and at flow rates that resulted in column residence times in the range of 45–325 min. Unusual, but similar, breakthrough curves were observed in each of the surfactant transport studies. Figure 6 shows such a breakthrough curve for an influent surfactant concentration of 50 times the CMC (1.8×10^{-4} mol/L) (*11*). The breakthrough curve has been normalized to a non-dimensional form to depict relative surfactant concentrations (C_s/C_o) plotted against pore volumes of aqueous solution flushed through the column, where C_s is the effluent surfactant concentration and C_o is the influent surfactant concentration.

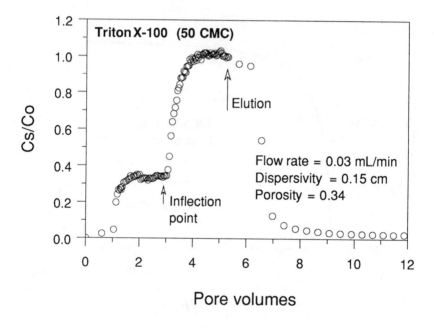

Figure 6. Surfactant breakthrough and elution curves for transport of Triton X-100 through a column packed with Lincoln fine sand.

The first breakthrough of surfactant was observed after flushing slightly more than one pore volume. This was followed by a plateau in the effluent aqueous surfactant concentration which persisted up to an "inflection point". The inflection point on the breakthrough curve was followed by a second-stage breakthrough curve with tailing of the effluent concentration as it approached the influent concentration value, C_o. The elution curves did not show a stage-wise behavior as was evident for the breakthrough curves and considerable tailing was observed. The average surface concentration of the surfactant at the inflection point can be computed by numerically integrating the area to the left of the breakthrough curve. The mass of sorbed surfactant is computed by multiplying this area by the influent surfactant concentration, and subtracting one pore volume representing the volume of water in the column prior to pumping the surfactant solution. This corrected mass of sorbed surfactant can then be divided by the total surface area of the sand to estimate an average surface concentration.

The computed surface coverage at the inflection point on the breakthrough curve for the data shown in Figure 6 was 180 $Å^2$/molecule. Similar values were computed for other breakthrough curves not shown. This value of surface coverage is very similar to the surface coverage corresponding to C_{int} shown in Figure 1. Thus, it is proposed that the inflection point reflects a change in the sorption regime from a type of patchy monolayer to partial bilayers. Based on equilibrium sorption experiments, several researchers have proposed that admicelles or partial bilayers are created on silica surfaces (*6, 10, 19*), as well as Lincoln fine sand (*5*).

These observed surfactant transport phenomena are significantly different from those reported by Abdul and Gibson (*13*), who had conducted experiments with a nonionic alkyl ethoxylate surfactant (Witconol SN 70) in columns packed with a natural aquifer sediment, and who had observed breakthrough curves for the surfactant to be regular S-shaped curves. Abdul and Gibson (*13*) show that at 1% (v/v) surfactant influent concentration, the breakthrough curve shows a small degree of retardation. However, in their work, as the influent surfactant concentration was reduced, the effluent breakthrough curves became progressively more skewed; the number of pore volumes of surfactant solution required to reach breakthrough increased from 1.5 to more than 5 as the influent surfactant concentration was decreased from 1% to 0.125%. The work of Abdul and Gibson (*13*) indicates that the degree of retardation as well as the degree of skewness in surfactant breakthrough are dependent on influent surfactant concentrations. Magee et al. (*22*) have reported similarly retarded and skewed breakthrough curves for transport of natural dissolved organic matter through a dark quarry sand.

It may be expected that the retardation in surfactant transport would occur as a result of sorption of the surfactant onto the solid matrix, while the skewness in the breakthrough curve would indicate kinetic or mass transfer limitations to surfactant transport. These considerations for the case of organic compounds have been discussed in a general fashion by Brusseau et al. (*23*). A sensitivity analysis by Brusseau et al. (*23*) has shown that the degree of skewness increases as the extent of non-equilibrium is increased for sorbing organic solutes. Some researchers have considered surfactant sorption phenomena in laboratory and field studies to be either equilibrium processes or to be negligible altogether (*24, 25*).

Sabatini and Austin (*14*) have reported "two-leg breakthrough curves" for transport of rhodamine WT through alluvial aquifer sand. A variation in pore water velocity resulted in variation of rhodamine WT sorption in the second leg only, indicating that sorption in the first leg is not kinetically limited while that in the second leg is. Sabatini and Austin (*14*) did not determine the mechanisms respon-

sible for the "two-leg breakthrough" behavior. The results for Rhodamine WT
transport, however, are qualitatively similar to the observations for Triton X-100
transport.

Conceptual Model for Surfactant Transport.   As a first approximation,
an empirical, two-stage sorption-kinetic model is proposed for transport of Triton X-100 through Lincoln fine sand. This model envisions that at low, sub-
CMC concentrations, i.e., stage 1 of Figure 2, the surface coverage is sparse and
the maximum surface area for sorption of Triton X-100 onto the sand may be
around 700-800 Å² per molecule, as determined from Figure 1 (5). The surfactant
molecules at such surface coverage may be visualized as more-or-less flat-lying
on the surface of the solid. As the concentration of the surfactant is increased
such that the aqueous-phase concentrations are in excess of the CMC, a much
greater surface coverage is observed, corresponding to stage 2 of Figure 2 with the
progressive formation of admicelles or bilayers.

This categorization of sorption behavior has been proposed by several researchers studying equilibrium sorption of nonionic surfactant onto silica gels,
silica sols, and sands. Clunie and Ingram (19) have proposed five stages of nonionic surfactant sorption onto a solid surface, where the sorbed molecules transform from a sparse flat-lying coverage to a denser monolayer, followed by the
formation of multilayers and surface clusters as the surfactant concentration is
increased. Partyka et al. (6) have described a similar sorption process in three
distinct stages. Levitz and van Damme (10) have also explored the possibility of
formation of close-packed assemblies on the surface that presumably act as precursors to the micellization process. All of these studies point towards two broad
categories of sorption phenomena: monomer-surface interactions and monomer-
monomer interactions. This concept forms the basis of the proposed two-stage,
sorption kinetic model.

Mathematical Formulation.   An empirical mathematical model is formulated
in order to interpret Triton X-100 transport phenomena observed in sand column
tests. The model comprises two first-order sorption reactions which work sequentially, i.e., the first reaction is valid up to the inflection point on the breakthrough
curve, where surface coverage is less than 2.4 μmol/g, and the second reaction is
valid only for higher surface coverage. Mathematically,

$$C_s \rightleftharpoons^{k_{f1}}_{k_{b1}} S_1 \tag{8}$$

$$C_s \rightleftharpoons^{k_{f2}}_{k_{b2}} S_2 \tag{9}$$

where k_f and k_b are the first-order forward and reverse (backward) reaction constants for the sorption reaction, and the subscripts 1 and 2 indicate the controlling sorption regime, i.e., stage 1 or 2, and the total sorbed surfactant mass at
equilibrium is $S_s = g(C_s) = S_1 + S_2$. Each sorption regime can be defined by
a corresponding governing differential equation (the detailed derivation of these
equations is discussed elsewhere (11)):

$$\frac{dC_s}{dt} = -k_{f1}C_s \tag{10}$$

$$\frac{dC_s}{dt} = \frac{-k_{b2}\rho}{n}[g(C_s) - S_1 - S_2] \tag{11}$$

Figure 7 presents the predictions from the model as fitted to the experimental data from Figures 6. The two-stage kinetic model appears to capture the essential features of the breakthrough curves including the plateau region and the prolonged tailing. A sensitivity analysis was performed to evaluate the dependence of model output on the two fitting parameters, k_{f1} and k_{b2}. It was observed that the first-stage reaction, represented by k_{f1}, was much faster than the second-stage reaction, represented by k_{b2}. This observation qualitatively agrees with the experimental evidence for two-leg breakthrough of rhodamine WT in aquifer sediment (*14*). The effect of pore-water velocities on these fitting parameters is not discernible from the available data (*11*).

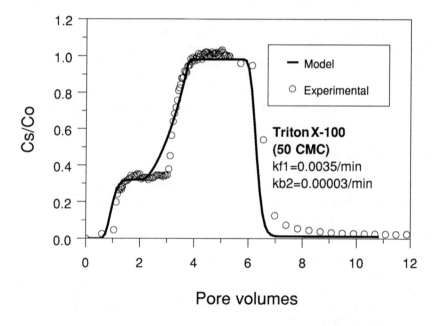

Figure 7. Comparison of model predictions obtained from fitting k_{f1} and k_{b2} to the experimental data shown in Figure 6 from transport of Triton X-100 through Lincoln fine sand.

Implications of Surfactant Sorption and Transport Kinetics

The column and batch experiments have provided insights into various surfactant sorption phenomena. It is observed that Triton X-100 nonionic surfactant sorption onto aquifer sediments is not an equilibrium process, but is rate-limited and comprises different sorption regimes depending on the surfactant concentration. A first regime may be envisioned as having a relatively sparse surface coverage for which sorption is comparatively rapid, while the second regime entails the formation of surfactant clusters, such as partial bilayers, for which sorption is relatively slower. The two-stage kinetic sorption model was used to fit the data from additional column transport experiments for Triton X-100. In general, the modeling approach captured the salient features of the experimental data. The fitted values of k_{f1} and k_{b2} varied significantly with influent surfactant concentration and flow rate; a much stronger dependence on influent concentration was observed compared to flow rate. An explanation of this variance cannot be provided at present.

This work indicates the need for development of techniques to observe and to ascertain relevant sorption phenomena at a molecular-scale. With respect to possible applications for *in-situ* surfactant-enhanced remediation strategies, the retardation in surfactant transport due to rate-limited sorption phenomena should be considered in assessing the duration of remediation, the appropriate surfactant concentration, and the total volume of surfactant solution required. The removal of the nonionic surfactant from sand columns by flushing also appears to be a slow, rate-limited process and long time periods may be required for complete removal of the sorbed surfactant by flushing. This means that a residual sorbed surfactant mass may be left behind in practical surfactant-aided remediation schemes.

Acknowledgments

This research was supported by the US Environmental Protection Agency, Office of Exploratory Research, Washington, DC, through Grant Number R819266-01-0.

Literature Cited

(1) Vigon, B.W.; Rubin, A.J. *J. Water Pollut. Control Fed.* **1989**, *61*, 1233-1240.
(2) Abdul, A.S.; Gibson, T.L.; Rai, D.N. *Ground Water* **1990**, *28*, 920-926.
(3) Fountain, J.C.; Klimek, A.; Beikirch, M.G.; Middleton, T.M., *J. Haz. Mater.* **1991**, *28*, 295-311.
(4) Liu, Z.; Edwards, D.A.; Luthy, R.G. *Water Res.* **1992**, *26*, 1337-1345.
(5) Edwards, D.A.; Adeel Z.; Luthy, R.G. *Environ. Sci. Technol.* **1994**, *28*, 1550-1560.
(6) Partyka, S.; Zaini, S.; Lindheimer, M.; B. Brun, B. *Colloids and Surfaces* **1984**, *12*, 255-270.
(7) Rupprecht, H.; Gu, T. *Colloid Polymer.. Sci.* **1991**, *269*, 506-522.
(8) Gu, T.; Zhu, B. *Colloids and Surfaces* **1990**, *44*, 81-87.

(9)   Somasundaran, P.; Snell, E.D; Xu, Q. *J. Colloid and Interface Sci.* **1991**, *144*, 165-173.
(10) Levitz, P.; van Damme, H. *J. Phys. Chem.* **1986**, *90*, 1302-1310.
(11) Adeel, Z.; Luthy, R.G. *Environ. Sci. Technol.*, submitted.
(12) Bohmer, M.; Koopal, L. *Langmuir* **1990**, *6*, 1478.
(13) Abdul, S.A.; Gibson, T.L. *Environ. Sci. Technol.* **1991**, *25*, 665-671.
(14) Sabatini, D.A.; Austin, T.A. *Ground Water* **1991**, *29*, 341-349.
(15) Nelson, D.W.; Sommers, L.E. In *Methods of Soil Analysis, Part 2, Chemical and Microbiological Properties*; Page, A.L., Ed.; American Society of Agronomy, Inc.: Madison, Wisconsin, **1986**.
(16) Wilson, J.T.; Enfield, C.G.; Dunlap, W.T.; Cosby, R.L.; Foster, D.A.; Baskin, L.B. *J. Environ. Qual.* **1981**, *10*, 501-506.
(17) Gregg, S.J.; Sing, K.S.W. *Adsorption Surfaces Area and Porosity*; Academic Press: London, **1967**.
(18) Robson, R.J.; Dennis, E.A. *J. Phys. Chem.* **1977**, *81*, 1075-1078.
(19) Clunie, J.S.; Ingram, B.T. In *Adsorption from Solution at the Solid/Liquid Interface*; Parfitt, G.D., Rochester, C.H., Eds.; Academic Press: New York, New York, **1983**.
(20) Means, J.C.; Wood, S.G.; Hassett, J.J.; Banwart, W.L. *Environ. Sci. Technol.* **1980**, *14*, 1524-1528.
(21) Edwards, D.A.; Liu, Z.; Luthy, R.G. *J. Env. Eng., ASCE* **1992**, *120*, 5-22.
(22) Magee, B.R.; Lion, L.W.; Lemley, A.T. *Environ. Sci. Technol.* **1991**, *25*, 323-331.
(23) Brusseau, M.L.; Jessup, R.E.; Rao, P.S.C. *Water Resour. Res.* **1989**, *25*, 1971-1988.
(24) Abriola, L.M.; Dekker, T.J; Pennell, K.D. *Environ. Sci. Technol.* **1993**, *27*, 2341-2351.
(25) Thurman, E.M.; Barber, L.B. Jr.; LeBlanc, D. *J. Contam. Hydrol.* **1986**, *1*, 143-161.

RECEIVED December 13, 1994

Chapter 5

Sorption of Nonpolar Organic Compounds, Inorganic Cations, and Inorganic Oxyanions by Surfactant-Modified Zeolites

Robert S. Bowman, Grace M. Haggerty, Roger G. Huddleston, Daphne Neel, and Matthew M. Flynn

Department of Geoscience and Geophysical Research Center, New Mexico Institute of Mining and Technology, Socorro, NM 87801

Treatment of natural zeolites with cationic surfactants yielded sorbents with a strong affinity for nonpolar organics and for inorganic oxyanions, and caused little decrease in the zeolite's sorption of transition metal cations. Two zeolites modified with hexadecyltrimethylammonium or methyl-4-phenylpyridinium remained chemically stable in aggressive aqueous solutions and organic solvents. The modified zeolites sorbed benzene, toluene, p-xylene, ethylbenzene, 1,1,1-trichloroethane, and perchloroethylene from aqueous solution via a partitioning mechanism; sorption affinity was in the order of the sorbates' octanol-water partition coefficients. Zeolites with or without surfactant treatment strongly sorbed Pb^{2+} from solution. Surfactant-modified zeolite also sorbed chromate, selenate, and sulfate from solution; the mechanism appears to be surface precipitation of a surfactant-oxyanion complex.

Zeolites are hydrated aluminosilicate minerals characterized by cage-like structures, high internal and external surface areas, and high cation exchange capacities. Both natural and synthetic zeolites find use in industry as sorbents, soil amendments, ion exchangers, molecular sieves, and catalysts. Clinoptilolite is the most abundant naturally occurring zeolite. It has a two-dimensional 8-ring and 10-ring channel structure with the largest aperture measuring 4.4 by 7.2 Å (1). The unit cell formula is $(Ca,Na_2,K_2)_3[Al_6Si_{30}O_{72}] \cdot 24H_2O$. Most zeolites used in the water treatment field are not true zeolites but are amorphous sodium aluminosilicates with limited cation exchange capacities. The low cost of natural zeolites ($60-$100/ton) makes their use attractive in pollution abatement applications.

    Zeolite chemistry resembles that of smectite clays. In contrast to clays, however, natural zeolites can occur as mm- or greater-sized particles and are free of shrink/swell behavior. As a result, zeolites exhibit superior hydraulic characteristics and are suitable for use in filtration systems (2) and potentially as

0097–6156/95/0594–0054$12.00/0
© 1995 American Chemical Society

permeable barriers to slow or prevent dissolved chemical migration in groundwater (Figure 1). Due to their large surface areas and cation exchange capacities, natural zeolites have a high affinity for the cationic transition metals such as Pb^{2+} and Cd^{2+}. These materials have been used commercially to remove Pb^{2+} and NH_4^+ from wastewaters (*3,4*). Natural zeolites have little affinity for inorganic anions such as chromate or for nonpolar organics like those found in fuels or solvents.

Treatment of natural zeolites with cationic surfactants dramatically alters their surface chemistry (*5*). These large organic cations exchange with native cations such as Na^+, K^+, or Ca^{2+} on the external zeolite surfaces (Figure 2). If the surfactant is larger than the largest aperture of the zeolite, inorganic cations on internal surfaces will not be displaced. The surfactant-modified zeolites gain an affinity for nonpolar organics and inorganic anions, while retaining much of their sorption capacity for transition metal cations.

This paper summarizes mostly unpublished data on the sorption of nonpolar organics, inorganic cations, and inorganic oxyanions by surfactant-modified zeolites, and discusses the use of these materials in environmental applications. More detail on the work described here can be found in the references *6-9*.

Materials and Methods

Zeolite Sources and Properties. Two different natural zeolites were used for this work. The Tilden zeolite was supplied by the Zeotech Division of Leonard Minerals from their mine in Tilden, Texas. The St. Cloud zeolite was obtained from the St. Cloud mine near Winston, New Mexico. The mineralogical content of each zeolite was determined by X-ray diffraction. The internal and external cation exchange capacities (CECs) were determined using the method of Ming and Dixon (*10*). The properties of the two zeolites are summarized in Table I. Both zeolites are of the clinoptilolite type. The St. Cloud zeolite is a "purer" zeolite in the sense that it is free of other minerals possessing significant surface areas or CECs. Both zeolites have about the same total CEC (50-60 me/100 g), but the Tildon sample has a higher fraction of external CEC to internal CEC. The greater apparent external CEC of the Tildon sample is probably due to the large fraction of smectite in this sample. The interlayer space of smectite clays would be measured as external CEC by the method of Ming and Dixon (*10*).

Surfactants. Two different surfactants were used to treat the zeolites. Hexadecyltrimethylammonium bromide or chloride (HDTMA) was obtained from Aldrich Chemical Company, Inc. (Milwaukee, WI). Methyl-4-phenylpyridinium (MPP) was synthesized in our laboratory as described by Huddleston (*6*). Carbon-14 methyl-labelled HDTMA and MPP were obtained from American Radiolabelled Chemicals, Inc. (St. Louis, MI).

Preparation of Surfactant-Modified Zeolite. The zeolite, ground and sieved to a size range of 0.15-2.0 mm, was treated with a quantity of HDTMA or MPP equal to 100% of the external CEC. Typically, 40 g of zeolite and 120 mL of aqueous surfactant solution of the appropriate concentration were placed in a 250-mL centrifuge bottle. The bottles were mechanically shaken for 24 hours,

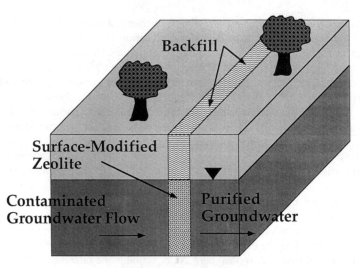

Figure 1. Potential use of surfactant-modified zeolites as permeable barriers.

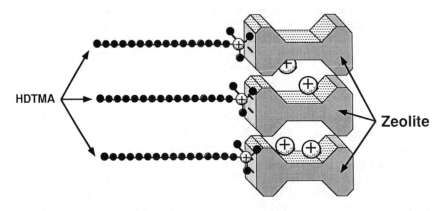

Figure 2. Schematic diagram of the cationic surfactant
hexadecyltrimethylammonium bound to a zeolite's external cation exchange
sites, with internal exchange sites occupied by inorganic cations.

Table I. Properties of the Two Zeolites Tested

	Mineralogy (%)					CEC (me/100g)	
Zeolite	Clinoptilolite	Smectite	Quartz	Amorphous	Lithic minerals	external	internal
Tilden	60	20	0	15	5	30	30
St. Cloud	60	0	20	0	20	15	35

centrifuged, and decanted. The surfactant-modified zeolite was then rinsed with distilled water several times before air-drying and storage.

Preliminary experiments with ^{14}C-labelled HDTMA or MPP showed that surfactant was exchanged stoichiometrically up to the zeolites' external CECs. This resulted in organic carbon fractions of 0.063 kg/kg for HDTMA-Tilden and 0.038 kg/kg for MPP-Tilden, and of 0.034 kg/kg for HDTMA-St. Cloud. The St. Cloud zeolite was not treated with MPP. Carbon oxidation of the treated zeolites yielded organic carbon fractions equal to 90-100% of those determined by [14]C (*6*).

Stability of Organo-Zeolite. The strength of the surfactant-zeolite association was tested by attempting to wash off the surfactant with a variety of aqueous solutions and organic solvents. The ^{14}C-labeled HDTMA- or MPP-treated zeolite was subjected to buffered solutions of pH 3, 5, and 10. The buffers consisted of 0.05 M potassium hydrogen phthalate-0.02 M hydrogen chloride (pH 3), 0.1 M sodium acetate - acetic acid (pH 5) and 0.025 M sodium bicarbonate - 0.01 M sodium hydroxide (pH 10). The surfactant-modified zeolite was also subjected to extremes in ionic strength (1.0 M $CaCl_2$ and 1.0 M CsCl), and to organic solvents (benzene, toluene, methanol). We also attempted to wash bound surfactant off of the zeolite using solutions of other surfactants including tetramethylammonium (TMA) and phenyltrimethylammonium (PTMA). Twenty milliliters of each of these solutions was added to Teflon centrifuge tubes containing 2.5 g of the radiolabelled organo-zeolite and continuous shaking begun at 25°C. Duplicate samples and appropriate blanks were prepared for each solution concentration. Aliquots of the solutions were removed at intervals over a 72-h period and analyzed for ^{14}C-labelled surfactant.

Organic Chemical Sorption. Batch sorption experiments were performed using a background electrolyte of 0.005 M $CaCl_2$. Aqueous solutions of benzene, toluene, *p*-xylene, ethylbenzene, 1,1,1-trichloroethane (TCA), or perchloroethylene (PCA) were prepared at five or six initial concentrations up to about half of their aqueous solubilities. Solutions were injected into 15-mL crimp-top vials with Teflon-lined seals containing 2.5 g of the surfactant-modified zeolite. An attempt was made to expel all air from the vials. The zeolite-solution mixtures were placed on a shaker at 25° C for 24 hr, a period shown sufficient to attain sorption equilibrium. After equilibration, approximately 2 mL was withdrawn from each vial using a gas-tight syringe and placed in gas chromatography (GC) autosampler vials, again attempting to eliminate any air in the vial. Appropriate blanks prepared with and without zeolite received the same treatment. Samples with untreated zeolite (no surfactant) were also prepared for each initial condition. All samples and blanks were prepared in duplicate or triplicate. Equilibrium solutions were analyzed on a Hewlett-Packard 5890 GC equipped with a flame ionization detector and a 10-m HP-5 (5% phenyl methylsilicone) capillary column. Further details on the analytical methods are provided in Huddleston (*6*) and Neel (*7*). Sorption of the organics was calculated from the difference between initial and final concentrations, making appropriate corrections for the blanks.

Table II. Percent of Surfactant Remaining on the Zeolite Surface after 72-Hour Exposure to Various Solutions/Solvents

Solution/Solvent	St. Cloud Zeolite % Surfactant Remaining HDTMA	Tilden Zeolite % Surfactant Remaining HDTMA	MPP
distilled water	99.3	99.5	99.5
0.005 M CaCl$_2$	99.1	99.6	99.0
pH 3	98.3	99.7	98.6
pH 5	98.3	ND[†]	ND
pH 10	99.0	99.6	99.0
0.10 M CsCl	98.6	99.6	97.0
1.0 M CsCl	97.2	99.1	90.4
50 mg/L CrO$_4^{2-}$	99.4	ND	ND
methanol	96.0	91.2	99.8
benzene	99.6	99.8	99.9
toluene	99.6	ND	ND
0.15M HDTMA	--	--	76.6
0.15M MPP	ND	98.7	--
0.10M TMA[‡]	ND	98.9	94.2
1.0 M TMA	ND	97.7	85.7
0.1 M PTMA[§]	ND	99.3	98.3
1.0 M PTMA	ND	68.0	68.7

[†] ND - not determined
[‡] TMA - tetramethylammonium
[§] PTMA - phenyltrimethylammonium

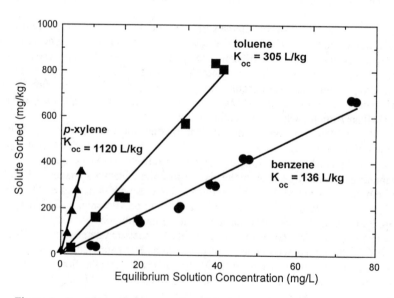

Figure 3. Sorption of benzene, toluene, and p-xylene by HDTMA-treated Tilden zeolite.

Lead Sorption. Lead sorption isotherms were prepared in a manner similar to that for the organic solutes. Solutions of Pb^{2+} in the form of $PbCl_2$ were prepared for initial Pb concentrations of 0-100 mg/L in 0.005 M $CaCl_2$. Ten milliliters of Pb solution was added to 2.5 g of zeolite in a 50-mL polyallomer centrifuge tube. The tube was shaken for 24 hr at 25 °C, centrifuged, and the supernatant analyzed for Pb concentration by atomic absorption spectroscopy. Sorption of Pb was calculated by difference.

Anion Sorption. Anion solutions were prepared using potassium chromate, sodium selenate, or sodium sulfate dissolved in 0.005 M $CaCl_2$. Initial anion concentrations ranged from 5 mg/L to 200 mg/L. Two and one-half grams of zeolite and 10 mL of anion solution were placed in a 50-mL centrifuge tube. Duplicate samples and appropriate blanks were prepared for each solution concentration. The tubes were mechanically shaken for 24 h at 25 °C, a period shown sufficient for attaining sorption equilibrium. Each sample was centrifuged and 5 mL of the supernatant was decanted for analysis. Total chromium and total selenium were determined by atomic absorption, while sulfate was analyzed using ion chromatography. The amount of anion sorbed was determined from the difference in solution concentration before and after equilibration.

Results and Discussion

Stability of Surfactant-Modified Zeolite. The surfactant-modified zeolite was stable when immersed in a variety of aggressive solutions (Table II). Extremes in solution pH, ionic strength, or organic solvents failed to remove surface-bound surfactant. Even a 1.0 M solution of Cs^+, which is strong competitor for cation exchange sites, was unable to displace a significant amount of HDTMA. The only solutes which could displace HDTMA or MPP from the zeolite were high concentrations (1.0 M) of other high molecular weight cationic surfactants such as PTMA. Similarly, at a concentration of 0.15 M, HDTMA could displace MPP from the Tilden zeolite. This strong selectivity of mineral surfaces for cationic surfactants has been noted by others (*11*). The surfactant's affinity for the zeolite surface may be due to a combination of van der Waals associations among the sorbed surfactant hydrocarbon chains and entropic effects (*11,12*).

Organic Chemical Sorption. Sorption isotherms for benzene, toluene, and *p*-xylene on HDTMA-modified Tilden zeolite are shown in Figure 3. All of these solutes were strongly-retained by the surfactant-modified zeolite; the untreated zeolite showed no affinity for the organics. The organic-carbon based linear sorption coefficients (K_{oc}) for the three solutes are also shown in Figure 3. The K_{oc}s were calculated based on an organic carbon content of 6.3% by weight for Tilden zeolite treated with HDTMA to 100% of the external CEC. A number of characteristics of nonpolar organic sorption to surfactant-modified zeolite is shown by Figure 3. First, all of the isotherms are linear. Second, sorption increases as the aqueous solubility of the organic decreases from benzene (1791 mg L^{-1}) to toluene 535(mg L^{-1}) to *p*-xylene (200 mg L^{-1}). Both of these characteristics are consistent

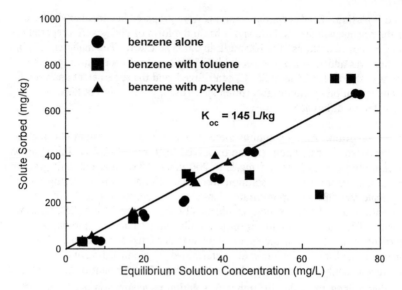

Figure 4. Benzene sorption by HDTMA-treated Tilden zeolite in the presence of equal amounts of toluene or *p*-xylene.

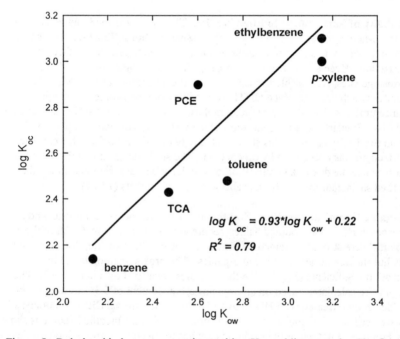

Figure 5. Relationship between experimental log K_{oc} and literature log K_{ow} for sorption of nonpolar organics by HDTMA-treated Tilden zeolite.

with a partitioning-type retention mechanism (*14*). Additional evidence for a partitioning mechanism is the lack of competition when multiple nonpolar sorbates are present. Figure 4 shows the results for the sorption of benzene alone and in the presence of equal initial concentrations of toluene or *p*-xylene. The K_{oc} for benzene changes by only about seven percent (from 136 to 145 L kg^{-1}) when potentially competing solutes are present. Parallel experiments for toluene and *p*-xylene sorption in the presence of the other two nonpolar organics showed a similar lack of competition (*7*).

Sorption of ethylbenzene, TCA, and PCA on HDTMA-Tilden zeolite or MPP-Tilden zeolite also yielded linear isotherms (*6*). There was a strong correlation between the log K_{oc} and the log of the octanol-water partition coefficient (log K_{ow}) for each of the six organic sorbates tested (Figure 5). This correlation between K_{oc} and K_{ow} is further evidence for a simple partitioning of nonpolar organics from aqueous solution onto the surfactant-modified zeolite. Log K_{oc} values for sorption of organics by MPP-zeolite were similar to those for HDTMA-zeolite (*6*), indicating that sorption was controlled primarily by the surface organic carbon content and only secondarily by the nature of the surfactant "tail" (aliphatic versus aromatic in the case of HDTMA and MPP).

Partitioning has also been used to explain the sorption of nonpolar organics by surfactant-modified clays (*15*). Unlike the smectite clays, however, surfactant-modified zeolite does not show selectivity for the six nonpolar organics tested beyond their relative hydrophobicities as expressed by their log K_{ow}s. In contrast, due to retention of surfactant on their interlayer spaces, expanding clays can show additional selectivity based on the molecular size of the sorbate (*16*). Depending upon the application, this additional selectivity could be an advantage or disadvantage.

Lead Sorption. As aluminosilicates with high specific surface areas and high cation exchange capacities, natural zeolites have a strong affinity for transition metal cations such as Pb^{2+}. Figure 6 shows sorption of Pb^{2+} by Tilden zeolite before and after treatment with HDTMA. Although the linear sorption coefficient (K_d) is lowered somewhat by the presence of the surfactant (from 37.2 to 25.6 L kg^{-1}), much of the sorption capacity for Pb^{2+} is retained. This can be understood by recalling that HDTMA and other large surfactant cations only occupy the zeolite's external cation exchange sites (Figure 2); internal exchange sites, accessible to small hydrated cations such as Pb^{2+}, are still available for sorption. In addition, Pb^{2+} and other transition metal cations can be retained by surface complexation reactions as well as via cation exchange. Thus, some sorption capacity should remain on the external surface of HDTMA-zeolite as well. The reduction of Pb^{2+} sorption by the zeolite after modification with HDTMA is likely due to a combination of loss of external cation exchange sites as well as blocking of some surface complexation sites by the relatively large, hydrophobic surfactant.

Anion Sorption. The surfactant-modified zeolites showed an unexpected affinity for the inorganic oxyanions. As crystalline solids with a net negative structural charge, natural zeolites, like smectite clays, tend to repel anions. Indeed, the untreated zeolites showed no sorption of chromate, selenate, or sulfate. Treatment

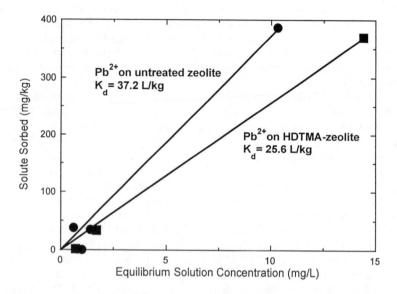

Figure 6. Sorption of Pb^{2+} by untreated and HDTMA-treated Tilden zeolite.

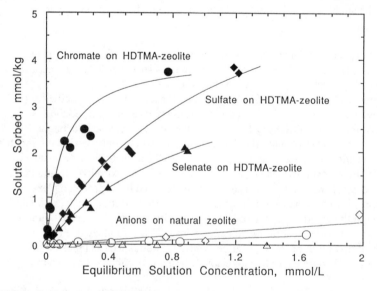

Figure 7. Sorption of chromate, selenate, and sulfate by HDTMA-treated St. Cloud zeolite. Reprinted from *Environ. Sci. Technol.* **1994**, 28, 452-458.

with HDTMA or other surfactants imparted a significant sorption capacity for these anions, however. The results for St. Cloud zeolite are shown in Figure 7. All of the anions displayed Langmuir-type sorption behavior on HDTMA-St. Cloud, with apparent sorption maxima ranging from 4-7 mmol kg^{-1} (5). Fourier-transform Raman spectroscopy of chromate sorbed on HDTMA-St. Cloud yielded a chromium signal shifted 10 cm^{-1} from the chromium signal of $(HDTMA)_2CrO_4$ solid. The magnitude of the Raman shift indicates strong influence of the zeolite surface on the chromium. This spectroscopic evidence, along with the good description of the sorption data by the Langmuir isotherm, suggest that the oxyanions are retained by a surface-precipitation mechanism (5). We have also found that the oxyanions arsenate and molybdate are retained by HDTMA-modified St. Cloud zeolite (data not presented).

Since all the anion isotherms were performed in a 0.005 M CaCl$_2$ matrix, Cl$^-$ did not compete effectively with the oxyanions for sorption. Indeed, sorption of chromate was only slightly reduced even at chloride levels as high as 1.0 M (9). The surfactant-modified zeolites thus may prove to be effective sorbents for toxic oxyanions even in contaminated waters having -high concentrations of monovalent anions such as Cl$^-$.

Summary and Conclusions

Surfactant-modified zeolites have been shown to be effective and versatile sorbents for nonpolar organics, inorganic cations, and inorganic oxyanions. Treatment of clinoptilolite-dominated zeolites with the high molecular weight cationic surfactants HDTMA or MPP yielded modified surfaces which were stable in aqueous solutions and in organic solvents. The surfactant-modified zeolites sorbed nonpolar organic solutes from water via a partitioning mechanism. The modified zeolite retained the high sorption capacity of the natural zeolite for Pb^{2+}. The surfactant-modified zeolite had an unexpectedly high affinity for inorganic oxyanions such as chromate, selenate, and sulfate. Anion retention appears to be due to a surface-precipitation mechanism.

The surfactant-modified zeolites thus exhibit the ability to sorb all the major classes of soil and groundwater contaminants. Preliminary experiments indicate that simultaneous sorption of nonpolar organics and inorganic cations or anions occurs with little competition among the different solute classes. The work presented here shows that the nonpolar organics do not compete with each other. The apparent mechanisms for anion and cation sorption suggest that multiple inorganic solutes of similar chemistry will compete with each other for sorption, but this remains to be tested.

Due to their broad sorptive capabilities, physical and chemical stability, and desirable hydraulic properties, surfactant-modified zeolites should find many applications in water pollution treatment and control. We estimate the cost of bulk quantities of these materials to be on the order of $0.30 to $0.50 per kilogram ($0.15 to $0.25 per pound). Surfactant-modified zeolite thus could be an attractive replacement for or addition to activated carbon or synthetic ion exchange resins in packed-bed treatment processes. Surfactant-modified zeolites seem particularly suited as permeable barriers for preventing groundwater pollutant migration, particularly where multiple contaminants are present. Their low cost and high permeabilities give them few if any competitors for this application.

Acknowledgments

This work was supported by the U.S. Geological Survey (Grant No. 14-08-0001-G1657) and the New Mexico Waste-management Education and Research Consortium (Project No. 91-41).

Literature Cited

(1) Newsom, J.M. *Science* **1986**, 231, 1093-109.
(2) Breck, D.W. *Zeolite Molecular Sieves: Structure, Chemistry, and Use*, John Wiley and Sons: New York, 1974; 771 p.
(3) Groffman, A.; Peterson, S.; Brookins, D. *Water Environ. Technol.* **1992**, May, 54.
(4) Mumpton, F.A.; Fishman. P.H. *J. Animal Sci.* **1977**, 45, 1188-1203.
(5) Haggerty, G.M.; Bowman, R.S. *Environ. Sci. Technol.* **1994**, 28, 452-458.
(6) Huddleston, R.G. *Surface-altered hydrophobic zeolites as sorbents for hazardous organic compounds*; Hydrology Open File Report No. H90-8; New Mexico Institute of Mining and Technology, Socorro, NM, 1990.
(7) Neel, D. *Quantification of BTX sorption to surface-altered zeolites*; Hydrology Open File Report No. H92-2; New Mexico Institute of Mining and Technology, Socorro, NM, 1992.
(8) Haggerty, G.M. *Sorption of inorganic oxyanions by organo-zeolite*; Hydrology Open File Report No. H93-1; New Mexico Institute of Mining and Technology, Socorro, NM, 1993.
(9) Flynn, M.M. *Sorption of chromate and lead onto surface-modified zeolites*; Hydrology Open File Report No. H93-9; New Mexico Institute of Mining and Technology, Socorro, NM, 1993.
(10) Ming, D.W.; Dixon, J.B. *Clays Clay Miner.* **1987**, 35, 463-468.
(11) Zhang, Z.Z.; Sparks, D.L.;Scrivner, N.C. *Environ. Sci. Technol.* **1993**, 27, 1625-1631.
(12) Vasant, E.F.; Uytterhoeven, J.B. *Clays Clay Miner.* **1972**, 20, 47-54.
(13) Cases, G.M.; Villieras, F. *Langmuir* **1992**, 8, 1251-1264.
(14) Chiou, C.T. In *Reactions and Movement of Organic Chemicals in Soils*; Sawhney, B.L.; Brown, K., Eds.; Soil Science Society of America: Madison, WI, 1989; pp
(15) Boyd, S.A.; Mortland, M.M.; Chiou, C.T. *Soil Sci. Soc. Am. J.* **1988**, 52, 652-657.
(16) Jaynes, W.F.; Boyd, S.A. *Soil Sci. Soc.. Am. J.* **1991**, 55, 43-48.

RECEIVED December 13, 1994

Chapter 6

Surfactant Selection for Optimizing Surfactant-Enhanced Subsurface Remediation

Bor-Jier Shiau[1], Joseph D. Rouse[1], David A. Sabatini[1,3,4], and Jeffrey H. Harwell[2,3]

[1]School of Civil Engineering and Environmental Science, [2]School of Chemical Engineering and Materials Science, and [3]Institute for Applied Surfactant Research, University of Oklahoma, Norman, OK 73019

Regulatory approval for surfactant enhanced subsurface remediation may be more readily achieved using food grade surfactants; results of solubilization and microemulsification studies using such surfactants are presented. For chlorinated organics (PCE, TCE and 1,2-DCE) solubility enhancements with food grade surfactants were one to two orders of magnitude relative to water alone via solubilization, with similar decreases in remediation times evidenced. Middle phase microemulsions (microemulsification) outperformed solubilization by an additional one to two orders of magnitude for the same surfactant concentration (up to four orders of magnitude more efficient than water alone). Microemulsification, however, was observed to be a function of surfactant structure, contaminant composition (including mixed DNAPL phases), and environmental conditions (e.g., aquifer temperature and hardness). Surfactant losses (precipitation, sorption) may hinder the technical and economical viability of the process. High performance surfactants with indirect food additive status (alkyl diphenyloxide disulfonates) were less susceptibility to losses (precipitation and sorption) than other ionic and nonionic surfactants while effectively solubilizing PAHs (i.e., naphthalene). It is thus observed that surfactant enhanced remediation can greatly expedite aquifer restoration and that surfactant selection is paramount to its technical and economical viability.

Chlorinated hydrocarbons are ubiquitous groundwater contaminants due to their widespread use as organic solvents and cleaners/degreasers. The immiscibility of chlorinated organics with groundwater results in their occurrence in the subsurface as residual (trapped) phases (thus the term dense nonaqueous phase liquids--DNAPLs). Water solubilities of these chlorinated hydrocarbons are frequently several orders of magnitude above their drinking water standards. Remediation of DNAPL residual saturation can require hundreds to thousands of pore volumes using conventional pump-and-treat methods; strongly sorbing (hydrophobic) compounds will experience a similar fate (e.g., polynuclear aromatic hydrocarbons and polychlorinated biphenyls;

[4]Corresponding author

0097–6156/95/0594–0065$12.00/0
© 1995 American Chemical Society

PAHs and PCBs, respectively). The inefficiency of conventional pump-and-treat methods for these contaminants has been addressed by several recent reviews; surfactants are mentioned in these reports as a promising technology for overcoming mass transfer limitations evidenced in conventional approaches (*1, 2*).

Two obstacles to widespread implementation of surfactant enhanced subsurface remediation will be (1) gaining regulatory approval for the injection of surfactants, and (2) the economics of the process, based largely on the capital costs of the surfactant (as discussed further in another chapter). Gaining regulatory approval is an obstacle common use of all chemical amendments. Surfactants with U.S. Food and Drug Administration direct food additive status are a focus of this chapter; these surfactants are common in food products and other consumer goods. The economics of surfactant-enhanced remediation processes will potentially be limited by surfactant losses in the subsurface (sorption, precipitation, etc.). Research investigating high performance surfactants that minimize these losses is also presented. Recovery and reuse of surfactants (which will also improve the economics) is the subject of another chapter.

Background

Surfactants are a class of compounds that are surface-active-agents. Surfactant molecules have two distinct regions (moieties); hydrophobic (water disliking) and hydrophilic (water liking). Thus, surfactant molecules migrate to interfaces where both portions of the molecule can be in a preferred phase. This causes surfactants to accumulate at air-water and oil-water interfaces, etc. Surfactants are common in detergents and in food products where their surface active nature is desirable. Above a certain concentration surfactant molecules self-assemble into aqueous phase spherical aggregates with the hydrophobic portions of the molecule in the interior of the aggregate and the hydrophilic portions at the exterior. This aggregate is referred to as a micelle, and the concentration above which micelles form is referred to as the critical micelle concentration (CMC). These micelles (aggregates) are highly soluble in water (due to their water-like exterior) but have an oil-like center or core. Thus, micelles can be considered as dispersed oil drops in the aqueous phase.

Surfactants can improve subsurface remediation by solubilization (increasing the aqueous concentration of the contaminant by partitioning into surfactant micelles) or microemulsification (formation of a middle phase microemulsion with concomitant ultra-low interfacial tensions). Solubilization enhancements result from partitioning of the contaminant into the oil-like core of the micelle, thereby effectively increasing the aqueous solubility of the contaminant and decreasing the number of water flushings required to extract the contaminant from the subsurface. Microemulsification enhancements result from the ultra-low interfacial tensions concomitant with formation of middle phase microemulsions. Significant reductions in the oil-water interfacial tension eliminates the capillary forces which cause the oil to be trapped, thereby allowing the oil to be readily flushed out with the water.

A middle phase system can be achieved by altering the surfactant system in several ways. In general, micellar systems transition from normal to swollen micelles (Winsor Type I), to middle phase systems (Winsor Type III), and finally to reverse micelles (Winsor Type II system, surfactants reside in the oil phase). It is possible that a mesophase (e.g., liquid crystal, coacervate) will result in the transitionary region rather than a middle phase system, thereby negating the potential benefits. However, measures can be taken to maximize the likelihood of achieving a middle phase system, as further discussed below. Figure 1 illustrates these various Winsor regions for a water-oil system containing equal volumes of water and oil. For a very hydrophilic surfactant system (right side of figure) the surfactant resides in the water phase as micelles and a portion of the oil phase partitions into the micellar phase. For a very

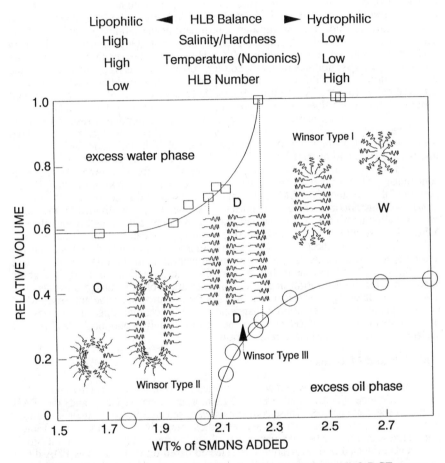

Figure 1: Example Phase Diagram Showing Winsor Systems (1,2-DCE versus SMDNS Concentration @ 15°C and 0.5 wt% AOT), Adapted from ref. 4.

lipophilic surfactant system (the left side of the figure) the surfactant resides in the oil phase as reverse micelles. Intermediate between these extremes a third phase appears; this new phase has a density intermediate between that of the oil and water phases and is comprised of water, oil and surfactant. The intermediate density of this new phase results in its designation as a middle phase microemulsion system. The interfacial tension reaches a minimum within this middle phase region (i.e., when equal volumes of oil and water occur in the middle phase system). Thus, theoretically the residual oil should flush out with the water (i.e., in several pore volumes).

These phase transitions can be realized by various methods, as denoted at the top of Figure 1. For ionic surfactants, adjustment of the salinity or hardness of the aqueous phase can produce the middle phase system, a strategy commonly utilized in surfactant enhanced oil recovery (3). However, introduction of high salt concentrations is not desirable in aquifer restoration, as remediation of brine contamination is also a difficult problem. In our research, middle phase systems are achieved by altering the hydrophilic - lipophilic balance (HLB) of a binary surfactant system; higher HLB values indicate increasing water solubility of the surfactant(s) (4).

Solubilization enhancement for neutral organic compounds results from the partitioning of the contaminant into the hydrophobic core of the micelle. Two parameters that describe this process are the molar solubilization ratio (MSR) and the micelle-water partition coefficient (K_m). The molar solubilization ratio (moles of contaminant per mole of surfactant) is determined from a graph of aqueous contaminant solubility versus surfactant concentration as the slope of the straight line portion of the plot above the CMC. The micelle-water partition coefficient (K_m) is the molar ratio (distribution) of the contaminant in the micellar phase divided by the molar ratio of the contaminant in the aqueous phase, and can be determined from values of MSR, water solubility of the contaminant, and molar concentration of water (4, 5).

Surfactant enhanced environmental remediation research to date has evaluated a wide spectrum of issues (space limitations prevent a more thorough listing: 5-16). This chapter will discuss the technical feasibility and limitations of using food grade surfactants for solubilization and microemulsification of chlorinated solvents, and describe the use of high performance surfactants for minimizing surfactant losses while not compromising remediation enhancement. This chapter thus emphasizes the importance of surfactant selection to the technical and economical feasibility of surfactant enhanced subsurface remediation.

Materials and Methods

The chlorinated organics evaluated in this research were tetrachloroethylene (PCE), trichloroethylene (TCE), and trans-1,2-dichloroethylene (DCE), and the PAH evaluated was naphthalene. These compounds were selected based on their ubiquitous occurrence as subsurface contaminants and their range in hydrophobicity. Table I summarizes characteristic parameters of these contaminants. The food grade surfactants evaluated in this research were selected based on their status as FDA direct food additive compounds and the HLB of the surfactants. The food grade surfactants are combinations of fatty acids and sugars. The S-MAZ surfactants and T-MAZ surfactants are sorbitan esters and ethoxylated sorbitan esters, respectively (with ethylene oxide groups ranging from 0 to 80). The high performance surfactants utilized are alkyl diphenyloxide disulfonates (DPDS) from the DOWFAX series, ranging from ten to sixteen carbons in the alkyl group (C10- to C16-DPDS). These surfactants have indirect food additive status from the USFDA. For comparison, a monosulfonate, sodium dodecylbenzene sulfonate (SDBS), was utilized. Table II summarizes properties of select surfactants discussed in this chapter. Surfactant solubilization, mobilization, precipitation and sorption studies were conducted according to standard procedures (see Shiau et al. (4, 18-20) and Rouse et al. (17, 21,

Table I: Contaminant Properties

Chemical	Molecular Formula	Molecular Weight	Aqueous Solubility (mg/L)	log K_{ow}[a]
Tetrachloroethylene (PCE)	C_2Cl_4	166	200	2.6
Trichloroethylene (TCE)	C_2HCl_3	131	1100	2.38
Trans-1,2-Dichloroethylene (DCE)	$C_2H_2Cl_2$	97	6300	0.48
Naphthalene	$C_{10}H_8$	128	32	3.3

[a]K_{ow} = octanol-water partition coefficient
After Rouse et al. (*17*) and Shiau et al. (*3*)

Table II: Surfactant Properties

Surfactant	MW[a]	Type	HLB[d]
SMDNS[b]	260	A	>40
POE (80) sorbitan monolaurate (T-MAZ 28)	3866	N	19.2
POE (20) sorbitan monolaurate (T-MAZ 20)	1266	N	16.7
POE (20) sorbitan monostearate (T-MAZ 60)	1310	N	14.9
Aerosol OT[c] (AOT)	445	A	?
Sorbitan Trioleate (S-MAZ 85K)	956	N	2.1
Sodium Dodecyl Benzene Sulfonate (SDBS)	348	A	
Sodium dodecyl diphenyloxide disulfonate (C12-DPDS)	575	A	
C16-DPDS (DOWFAX 8390)	642	A	

[a]MW = molecular weight
[b]Sodium mono and dimethyl naphthalene sulfonate
[c]Bis-2-ethylhexyl sodium sulfosuccinate
[d]HLB = hydrophilic-lipophilic balance; HLB values given for food grade surfactants only; high performance surfactants not evaluated for middle phase systems.
After Shiau et al. (*3*) and Rouse et al. (*17*)

22) for additional details on materials and methods). Sorption assays were conducted utilizing the Canadian River alluvium (CRA) material; it consists of 72% sand, and 27% silt and clay, and has an organic carbon content of 0.07%.

Food Grade Surfactants -- Solubilization / Microemulsification

Solubilization of chlorocarbons by single surfactant systems were conducted for SDS and three of the T-MAZ surfactants. Figure 2 summarizes the solubilization of the three chlorinated organics (PCE, TCE, 1,2-DCE) with T-MAZ 60 at 15 °C. It is observed that, as surfactant concentrations increase above the CMC, the aqueous phase concentration of the chlorinated organics increases linearly, in keeping with classical solubilization (partitioning) theory. These results illustrate several points: (1) below the CMC surfactant addition has little to no effect on the solubility of these contaminants, and (2) the higher the surfactant concentration is above the CMC (the greater the number of micelles) the greater is the chlorocarbon solubility enhancement (i.e., we will want to operate well above the CMC to achieve maximum enhancement, but below the surfactant's solubility limit to prevent phase separation). For example, from Figure 2 it is observed that for PCE the solubility enhancement is approximately two-fold at 10 mM and approximately eight-fold at 50 mM T-MAZ 60.

Table III: Solubilization Parameters for Chlorinated Organics

Chlorinated Organic	Surfactant	MSR	Log K_m
PCE	SDS	0.39	4.5
	T-MAZ 28	0.45	4.55
	T-MAZ 20	2.27	4.9
	T-MAZ 60	3.15	4.94
TCE	SDS	0.34	3.27
	T-MAZ 28	1.68	3.66
	T-MAZ 20	3.29	3.75
	T-MAZ 60	3.95	3.77
1,2-DCE	SDS	1.37	2.76
	T-MAZ 28	2.46	2.85
	T-MAZ 20	7.49	2.95
	T-MAZ 60	6.91	2.94

After Shiau et al., 1994a

Based on data in Figure 2, values of MSR and K_m for the chlorocarbons and T-MAZ 60 are summarized in Table III; also included in Table III are solubilization parameters for T-MAZ 20, T-MAZ 28, and SDS (data not shown). It is observed from Table III that the hydrophobic PCE evidences the greatest distribution into the micelles (highest K_m), and that K_m increases as the hydrophobicity of the contaminant increases (PCE > TCE > 1,2-DCE). While relatively minor deviations in K_m are observed between the surfactant types for a given contaminant, the data in Table III indicate that the distribution is more significantly impacted by the hydrophobicity of the contaminant (K_m varies more between contaminants for any surfactant than between surfactants for a given contaminant). This is reinforced by the results of West (1992), who observed similar K_m values for these chlorocarbons using alkylphenyl ethoxylated surfactants. Thus, for the solubilization mechanism, surfactant selection is relatively independent of the contaminant(s) and will most likely be made based on factors such as cost, susceptibility to losses, toxicity, etc. At the same time, when estimating (extrapolating) K_m values or when modeling the solubilization process, it should be noted that micellar solubilization (and thus K_m) varies as a function of contaminant type (nonpolar versus polar/ionic) and aqueous contaminant concentrations below the water solubility (*21, 23*). Aqueous contaminant concentrations below water solubility may be experienced due to nonequilibrium solubilization, mixed NAPL phases, etc. (*15, 21, 24*).

Our initial efforts to achieve middle phase microemulsions, without consideration of surfactant structure, were unsuccessful. The HLB of the surfactant systems was varied from 2.1 to 40.1; although phase inversion was observed (Type II to Type I) in this range, a clear middle phase was not achieved in the transition (instead a mesophase was realized). Using Aerosol OT (AOT) and sodium mono and di methyl naphthalene sulfonate (SMDNS) a middle phase microemulsion was realized in the transition region (as further discussed below). Thus, it is observed that surfactant type (structure) is critical to achieving microemulsification systems, unlike solubilization systems where enhancement is relatively independent of surfactant type.

Microemulsification with AOT and SMDNS was achieved by varying the SMDNS concentration while holding the AOT concentration constant. Figure 1 shows a phase diagram for 1, 2-DCE using AOT and SMDNS as the surfactant system. It is observed that at low SMDNS concentrations a Type II system is realized (the surfactant has partitioned into the oil phase). Increasing the SMDNS concentration enhances the surfactant balance (increases the affinity of the AOT for the interface), and results in a middle phase system. At yet higher SMDNS concentrations, the system is over-optimum and the surfactants reside in the water phase (Type I system). Thus, we observe that at intermediate SMDNS concentrations, the surfactant balance is achieved and a Winsor Type III (middle phase) system is realized. Middle phase systems were achieved for PCE, TCE, and 1,2-DCE individually (*4*) and in binary and ternary mixtures of these chlorocarbons (*19*) using this approach.

As an indication of the impact of ground water hardness on middle phase systems, Figure 3 shows middle phase systems for TCE at two levels of calcium (hardness). It is observed from Figure 3 that for higher hardness levels, more SMDNS is required to maintain the surfactant balance and achieve a middle phase system (the optimal SMDNS concentration is higher). This is expected as the increased calcium concentration will tend to drive the ionic AOT into the oil phase, and thus additional SMDNS is required to maintain the surfactant balance and retain the middle phase microemulsion. A similar response has been noted as a function of temperature (more SMDNS is necessary to achieve middle phase systems for lower temperature; *4*). Also, research has demonstrated that the surfactant composition necessary to achieve a middle phase microemulsion is affected by the composition of the residual phase. For example, the optimal SMDNS concentration (SMDNS*, which produces a minimum in interfacial tension within a given middle phase regime) for 0.5% AOT

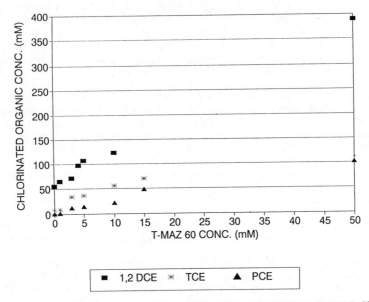

Figure 2: Solubilization of DNAPLs (PCE, TCE, 1,2-DCE) in T-MAZ 60 @ 15°C, Reproduced with permission from ref. *4*, Copyright 1994 Ground Water Publishing Co.

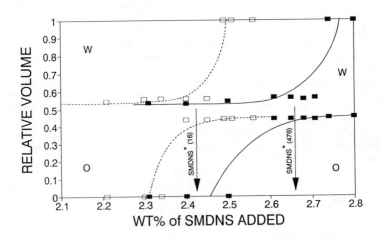

Figure 3: Phase Diagram for TCE at 15°C for AOT and SMDNS for Two Calcium Values (SMDNS* ( ) is the optimal SMDNS Concentration at Each Calcium Level), Adapted from ref. *4*.

with PCE, TCE, and 1,2-DCE are 1.40, 2.43 and 2.19 wt%, respectively (*4, 19*); this illustrates the impact of contaminant type on surfactant system design for microemulsification.

For mixed residual phases, the optimal surfactant concentration is a function of the mole fraction of each phase in the mixed waste and the optimal SMDNS concentration for each individual phase. For example, for 50% PCE and TCE the optimal SMDNS is approximately 2.1 wt% (*19*), intermediate between the optimal concentrations of each individual phase listed above (with minor deviations due to non-ideal mixing). Similar results have been observed for other binary and ternary chlorocarbon systems (*19*). These results illustrate the sensitivity of middle phase systems to aquifer conditions and contaminant composition, and should alert potential users to the likelihood of failure associated with utilizing middle phase microemulsions without proper surfactant selection and design efforts. Also, vertical migration of released residual may be realized and unacceptable, depending on aquifer conditions and the hydraulics of the extraction system. This again illustrates the care that must be taken in utilizing microemulsification systems. While recognizing these potential limitations, one must also remain cognizant of the fact that microemulsification has the potential to be significantly more efficient than solubilization (as further demonstrated below) and thus should not be prematurely dismissed as a viable technology.

In comparing the efficiency of the solubilization and microemulsification mechanisms, Table IV documents the enhancements of these two mechanisms for a common weight percent of surfactant via solubilization with T-MAZ 60 and microemulsification with Aerosol OT and SMDNS. As observed from Table IV, the enhancement is two orders of magnitude for PCE via solubilization and three and one-half orders of magnitude via microemulsification relative to water alone; for 1,2-DCE the enhancement by solubilization is approximately one order of magnitude, while being two orders of magnitude for microemulsification. The dramatic increase in efficiency via microemulsification versus solubilization is obvious. Table IV also demonstrates that enhancements for surfactant enhanced subsurface remediation will be greatest for the more hydrophobic compounds (via both solubilization and microemulsification).

Table IV: Comparison of Solubilization and Microemulsfication

Chlorinated Organic	GW Solubility (mg/L)	Solubilization (6.5 wt% T-MAZ 60)	Microemulsification (5.0 wt% AOT and SMDNS)
PCE	80.6	16900	619000
TCE	990	14700	594000
1,2-DCE	5340	37800	557000

After Shiau et al. (*4*)

Figure 4 compares the efficiency of solubilization and microemulsification based on one-dimensional column studies (*20*). Microemulsification achieved higher concentrations and eluted the PCE more quickly than solubilization (> 99% extracted in ca. 3 pore volumes). The tail on the solubilization curve indicates the reduced

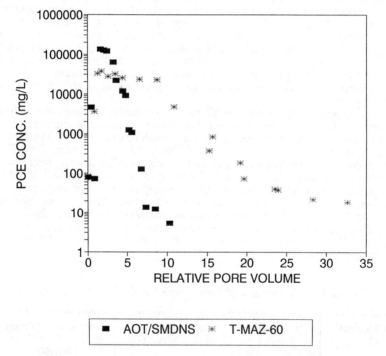

Figure 4: Column Results for PCE Elution via Solubilization (T-MAZ 60) and Microemulsification (AOT/SMDNS) Systems with CRA Medium.

extraction rate and thus slow approach to complete PCE elution for solubilization (likely due to interfacial area constraints); while ca. 85% of the PCE is eluted within 10 pore volumes, this value has not yet exceeded 90% by 30 pore volumes. Again, the potential advantages of microemulsification over solubilization are apparent.

It is thus observed that the food grade surfactants are viable for use in subsurface remediation activities. This is encouraging in light of obtaining regulatory approval for utilizing surfactants in subsurface remediation. Edible surfactants, however, may experience significant losses (due to precipitation, sorption, etc.), which may cause system failure (e.g., pore clogging due to precipitation or coacervation) and will result in increased costs of the surfactant (thereby hindering the economic viability of the process). For this reason, research in our laboratories is also evaluating the use of high performance surfactants (systems that are less susceptible to precipitation, sorption, etc.); this is the topic of the following section.

High Performance Surfactants -- Minimizing Losses

Surfactants were evaluated from the DOWFAX series based on the hypothesis that the disulfonated nature of these surfactants will reduce losses due to precipitation, sorption, etc. Results of precipitation assays conducted at 15°C are shown in Figure 5. For the SDBS assays, the presence of precipitate as a heavy white flock was easy to confirm by visual inspection in all cases. For C10-, C12- and C16-DPDS no precipitate was observed for calcium concentrations up to 0.1 M (log Ca(um) = 5), as confirmed with pinacyanol chloride assays. Thus, it is observed that the DOWFAX surfactants are significantly more resistant to precipitation than their monosulfonated equivalents (*17*). Table V summarizes K_{sp} values for various surfactants (including food grade surfactants discussed above, SDBS and DOWFAX 8390 (C16-DPDS)). Again, the significant decrease in precipitation by DOWFAX 8390 is noted by its much higher K_{sp} value. Similar reductions in precipitation have been observed for ethoxylated alkylsulfates (*22*). Also, cloud point measurements indicate that the nonionic surfactants should not phase separate under normal aquifer temperatures (*18*).

Table V: Surfactant Precipitation Constants (K_{sp} Values)

Surfactant	K_{sp} (@ oC)
Sodium dodecyl sulfate (SDS)	2.14e-10 (20) 3.72e-10 (25)
Sodium dodecyl benzene sulfonate (SDBS)	3.98e-10 (15) 8.40e-12 (25)
Linear alkyl diphenyloxide disulfonate (C16-DPDS)	> 1.0e-3 (15)
Aerosol OT (AOT)	2.38e-10 (15)
SMDNS	1.98e-08 (15)

After Shiau et al. (*18*)

Results of sorption experiments with SDBS and C12-DPDS are shown in Figure 6. It is common in soil sorption studies to utilize 0.01 N (0.005 M) Ca addition to

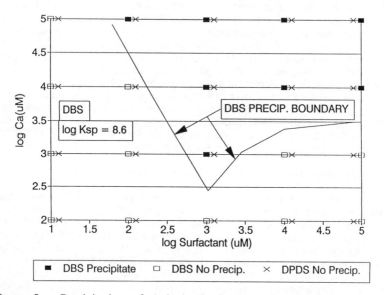

Figure 5: Precipitation of Anionic Surfactants with Calcium at 15°C, Reproduced with permission from ref. *17*, Copyright 1993 Environmental Science and Technology.

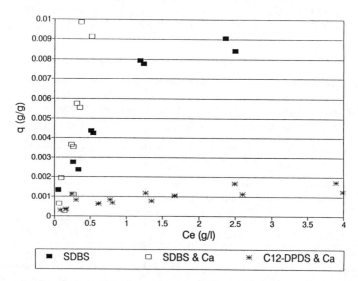

Figure 6: Sorption of SDBS and C12-DPDS with CRA (Calcium addition is 0.01N where noted), Reproduced with permission from ref. *17*, Copyright 1993 Environmental Science and Technology.

provide a uniform background matrix and promote the separation of solid and liquid phases for analysis. As shown in Figure 6, the SDBS isotherm with Ca yielded a higher slope than the SDBS assay without Ca, indicative of increased surfactant losses. In this case, however, the difference in slope is due to precipitation rather than sorption, as confirmed by inspection of Figure 5. Minimal precipitation is expected at low SDBS concentrations with 0.005 M Ca; however, with increases in the SDBS concentrations greater losses due to precipitation are anticipated (and thus greater deviations from the no-Ca-isotherm). Thus, the no-Ca isotherm is utilized to obtain the sorption parameters. These results illustrate the danger of confusing precipitation and sorption in experimental assays, and reinforces the danger precipitation poses to surfactant enhanced subsurface remediation (failure due to clogging, economics, etc.).

For comparison with SDBS, C12-DPDS was evaluated for its sorption on the same medium (as shown in Figure 6). Sorption isotherms conducted with and without Ca for the C12-DPDS demonstrated no deviation, reinforcing the lack of precipitation for the DPDS surfactants. Sorption parameters are summarized in Table VI; K_d values are applicable at low concentrations (linear region) and the Langmuirian coefficients (q_{max} -- maximum or plateau sorption value, and K_L) are applicable for all concentration values. For comparison purposes the sorption parameters for other surfactants are included in Table VI. From the isotherms in Figure 6 and the resulting sorption parameters in Table VI it is seen that SDBS is significantly more susceptible to sorption than the DOWFAX surfactants (K_d values three times greater for SDBS than C12-DPDS; q_{max} about seven times higher for SDBS than C12-DPDS). The sorption differences for nonionic surfactants evaluated in other research is equally if not more significant, approaching an order of magnitude in some cases (see Table VI). While AOT demonstrated low sorption potential, likely due to its twin tailed nature, its precipitation potential is concerning (see Table V). Thus, it is apparent that the

Table VI: Surfactant Sorption Losses With Subsurface Materials

Surfactant	K_d (cm³/g)	q_{max} (mg/g)	K_L
SDBS (C12)	8.3	11.4	0.729
C12-DPDS	3.1	1.6	0.703
T-MAZ 20	--	7	--
T-MAZ 80	--	6.6	--
AOT	--	1.8	--
CO 660*	19.7	--	--
CO 620*	41.1	--	--
Tergitol NP-10**	--	7.7	--
Triton X-100**	--	11.9	--

After Rouse et al. (*17*) and Shiau et al. (*18*)
* Alkylphenolethoxylate, same medium.
** Alkylphenolethoxylate, different medium.

disulfonated DOWFAX surfactants are significantly less susceptible to sorption losses than their monosulfonated equivalents and other surfactants as well. This, combined with their precipitation tolerance, make the DOWFAX surfactants especially attractive. Similar reductions in sorption have been observed for ethoxylated alkylsulfates (22).

The economic advantages of reducing surfactant sorption are obvious; there is also a technical advantage to reduced sorption. The sorbed surfactants, which can exist as a bilayer (admicelle) of surfactant molecules having a hydrophobic interior, can act as an organic sink for contaminants (much as organic matter does). While this phenomena can be exploited in sorbing barriers and unit separation processes (8, 23, 25), its occurrence in surfactant enhanced subsurface remediation can serve to delay aquifer restoration. This is especially true as elution of the adsorbed surfactant will decrease as the surfactant concentration declines (this can be assessed by the degree of nonlinearity of the sorption isotherm). Thus, again the advantage of low sorbing surfactants is apparent.

Having demonstrated that the DOWFAX surfactants are less susceptible to losses (precipitation and sorption) than their monosulfonated equivalents (as well as other surfactants), the obvious question is, How effective are these surfactants at enhancing contaminant remediation? Solubilization diagrams for naphthalene with SDBS, C10-, C12- and C16-DPDS are linear above the surfactant CMCs (17). Micellar-water partitioning coefficients (K_m) for naphthalene and the alkyl DPDSs were slightly higher than their monosulfonated equivalents (log K_m of 4.04 for SDBS and 4.32 for C12-DPDS, with C16-DPDS having a log K_m of 4.41). This indicates that the DOWFAX surfactants are equally if not more efficient in enhancing the solubility of contaminants relative to their monosulfonated equivalents. Thus, the DOWFAX surfactants appear very promising for use in subsurface remediation based on their ability to minimize surfactant losses while achieving high removal efficiency; this has been evidenced in tests ranging from batch and one-dimensional column studies to three-dimensional sand tank studies (17, 26). Again, this demonstrates the importance of surfactant selection for successful and economical implementation of surfactant enhanced subsurface remediation.

Acknowledgments

Although the research described in this article has been funded by the United States Environmental Protection Agency under assistance agreement No. CR 818553-01-0 and by the National Science Foundation under agreement BCS-9110780 to the University of Oklahoma, it has not been subjected to agency peer or administrative review and therefore may not necessarily reflect the views of the agency, and no official endorsement should be inferred.

Literature Cited

1.   Haley, J. L., Hanson, B., Enfield, C., and Glass, J. *Ground Water Monitoring Review*. Winter 1991, 119-124.
2.   Palmer, C. D. and Fish, W. "Chemical Enhancements to Pump and Treat Remediation," USEPA, EPA/540/S-92/001, 1992, 20 pp.
3.   Bourrel, M. and Schechter, R. S. *Microemulsions and Related Systems*. Surfactant Science Series, Vol. 30, Marcel Dekker, Inc., New York, 1988.
4.   Shiau, B. J., Sabatini, D. A. and Harwell, J. H. *Ground Water*. 32(4), 1994a, 561-569.
5.   Edwards, D. A., Luthy, R. G. and Liu, Z. *Environmental Science and Technology*. 25(1), 1991, 127-133.
6.   Valsaraj, K. T., and Thibodeaux, L .J. *Water Research*. 23(2), 1989, 183-189.

7. Vignon, B. W. and Rubin, A. J. *Journal of the Water Pollution Control Federation.* 61(7), 1989, 1233-1240.
8. Smith, J. A. and Jaffe, P. R. *Environmental Science and Technology.* 25, 1991, 2054-2058.
9. Jafvert, C. T. and Heath, J. K. *Environmental Science and Technology.* 25(6), 1991, 1031-1038.
10. Peters, R. W., Montemagno, C. D., Shem, L., and Lewis, B. A. *Hazardous Waste and Hazardous Materials.* 9(2), 1992, 113-136.
11. West, C. C., in *Transport and Remediation of Subsurface Contaminants: Colloidal, Interfacial and Surfactant Phenomena.* Sabatini, D. A. and Knox, R. C., eds., ACS Symposium Series 491, American Chemical Society, Washington, DC, 1992, 149-158.
12. West, C. C. and Harwell, J. H. *Environmental Science and Technology.* 26, 1992, 2324-2330.
13. Abdul, A. S., Gibson, T. L., Ang, C. C., Smith, J. C. and Sobczynski, R. E. *Ground Water,* 30(2), 1992, 219-231.
14. Fountain, J. C. in *Transport and Remediation of Subsurface Contaminants: Colloidal, Interfacial and Surfactant Phenomena.* Sabatini, D. A. and Knox, R. C., eds., ACS Symposium Series 491, American Chemical Society, Washington, DC, 1992, 182-191.
15. Pennel, K. D., Abriola, L. M. and Weber, W. J. Jr. *Environmental Science and Technology.* 27(12), 1993, 2341-2351.
16. Baran, J. R., Pope, G. A., Wade, W. H., Weerasooriyaa, V. and Yapa, A. *Environmental Science and Technology.* 28(7), 1994, 1361-1366.
17. Rouse, J. D., Sabatini, D. A. and Harwell, J. H. *Environmental Science and Technology.* 27(10), 1993, 2072-2078.
18. Shiau, B. J., Sabatini, D. A. and Harwell, J. H. "Properties of Food Grade Surfactants Affecting Subsurface Remediation of Chlorinated Solvents." 1994d, In Review.
19. Shiau, B. J., Sabatini, D. A. and Harwell, J. H. "Microemulsion of Mixed DNAPLs for Subsurface Remediation," 1994b, In Review.
20. Shiau, B. J., Sabatini, D. A. and Harwell, J. H. "Food Grade Surfactants for Subsurface Remediation: Column Studies." 1994c, In Review.
21. Rouse, J. D., Sabatini, D. A. and Harwell, J. H. "Effects of Unsaturated Hydrocarbon Concentrations on Anionic Surfactant-Enhanced Remediation using Semi-Equilibrium Dialysis," 1994a, In Review.
22. Rouse, J. D., Sabatini, D. A. and Harwell, J. H. "Minimizing Surfactant Losses Using Ethoxylated Alkylsulfates in Subsurface Remediation," 1994b, In Review.
23. Nayyar, S. P., Sabatini, D. A. and Harwell, J. H. "Surfactant Adsolubilization and Modified Admicellar Sorption of Nonpolar, Polar and Ionizable Organic Contaminants," *Environmental Science and Technology.* 28, 1994, 1874-1881.
24. Soerens, T. S., Sabatini, D. A., and Harwell, J. H. "Experimental and Modeling Studies of Surfactant Enhanced DNAPL Remediation in 1-D Columns." *Abstracts of 207th ACS National Meeting, American Chemical Society, San Diego, CA, March 13-17, 1994,* paper 151.
25. Sun, S. and Boyd, S. A. *Environmental Science and Technology.* 27, 1993, 1340-1346.
26. Roberts, B., Harwell, J. H., Sabatini, D. A., and Knox, R. C. "Sandtank Testing of Surfactant Enhanced Remediation of Subsurface Contamination by Chlorinated Solvents," Completion report submitted to Air Force Civil Engineering Support Agency, Civil Engineering Laboratory, Tyndall Air Force Base, Florida, April 1993.

RECEIVED January 23, 1995

BIOTIC–BIOSURFACTANT PROCESSES

Chapter 7

Biosurfactant- and Cosolvent-Enhanced Remediation of Contaminated Media

Mark L. Brusseau, Raina M. Miller, Yimin Zhang, Xiaojiang Wang, and Gui-Yun Bai

Soil and Water Science Department, University of Arizona, Tucson, AZ 85721

The use of water flushing (pump and treat, in situ soil washing) is one of the predominant methods currently in use for remediation of contaminated subsurface environments. While this method has been successful in some cases, its effectiveness is often constrained by one or more factors related to contaminant transport and fate. Recent research has focused on chemical additives that might be useful for enhancing contaminant removal during flushing. Examples include the addition of surfactants, cosolvents, and complexing agents. We are involved in the study of cosolvents and microbially produced surfactants (biosurfactants) and their effects on solubilization, biodegradation, and removal of residual phases from the subsurface. In this paper, we summarize our recent results and provide a comparison of the advantages and disadvantages in the use of biosurfactants and cosolvents in remediation of contaminated subsurface environments.

The use of water flushing (pump and treat, in situ soil washing) is one of the predominant methods currently in use for remediation of contaminated subsurface environments. While this method has been successful in some cases, its effectiveness is often constrained by one or more factors related to contaminant transport and fate. These factors include porous-media heterogeneity, dissolution of residual immiscible liquid, and rate-limited desorption (cf., 1-3). Recent research has focused on chemical additives that might be useful for enhancing contaminant removal during flushing. Examples include the addition of surfactants, cosolvents, and complexing agents. The basis for using chemical additives to enhance recovery of organic compounds

0097–6156/95/0594–0082$12.00/0

from porous media was established in petroleum science and engineering with the development of enhanced oil recovery (EOR) techniques. While the basic concepts developed for EOR are useful, it is not necessarily possible to apply them directly to environmental systems. Discussions of chemical enhancement techniques for environmental applications have been presented by several authors (3-7).

Surfactants are currently the focus of the research effort on chemical enhancements and, based on preliminary laboratory data, appear to have promise for enhancing pump-and-treat remediation in some situations. The use of dissolved organic matter (DOM) and of solvents is also being investigated, albeit at a smaller scale. Miscible solvents, such as ethanol, reduce the net polarity of the mixed solvent when added to water and thereby increase the quantity of a nonionic organic compound that can dissolve in the mixed solvent. This increase, in turn, results in a smaller equilibrium sorption constant and less attendant retardation. Thus, the addition of a cosolvent can reduce the volume of water required to flush a contaminant from porous media by altering the equilibrium phase distribution. A similar result is obtained with surfactants and DOM, although by different mechanisms. Hence, surfactants, DOM, and cosolvents act to increase the aqueous-phase concentration of organic compounds, the so-called "solubilization" effect. This effect is of special interest for the removal of residual phases of immiscible liquids and of highly sorbed solutes. The other major method of removing trapped residual phases is mobilization.

We are involved in the study of cosolvents and microbially produced surfactants (biosurfactants) and their effects on solubilization, biodegradation, and removal of residual phases from the subsurface. In this paper, we summarize our recent results and provide a comparison of the advantages and disadvantages in the use of biosurfactants and cosolvents in remediation of contaminated subsurface environments.

Biosurfactants: Rhamnolipids

Biosurfactants are a class of surfactants that are produced by microorganisms, plants, and animals (8). Of particular interest in remediation are bacterial biosurfactants, a structurally diverse class of anionic or nonionic compounds ranging from 500 to 1500 MW (7). Studies in our laboratory have shown that at least one type of biosurfactant, rhamnolipids, increases the apparent aqueous solubility of a variety of organic compounds (9-11). Rhamnolipids are produced by *Pseudomonas aeruginosa* strains, often in mixtures of several rhamnolipid types. Structure-function studies of rhamnolipids suggest that structural changes (Figure 1) caused differences in the mode of action and impact of the rhamnolipid on the apparent solubility of hexadecane (Figure 2). For example, the dirhamnolipid methyl ester acts primarily by emulsification, which explains its much greater impact on hexadecane concentration.

The monorhamnolipid acid has been tested for effectiveness of removal of residual non-aqueous phase liquids (NAPL). As shown in Figure 3 (from

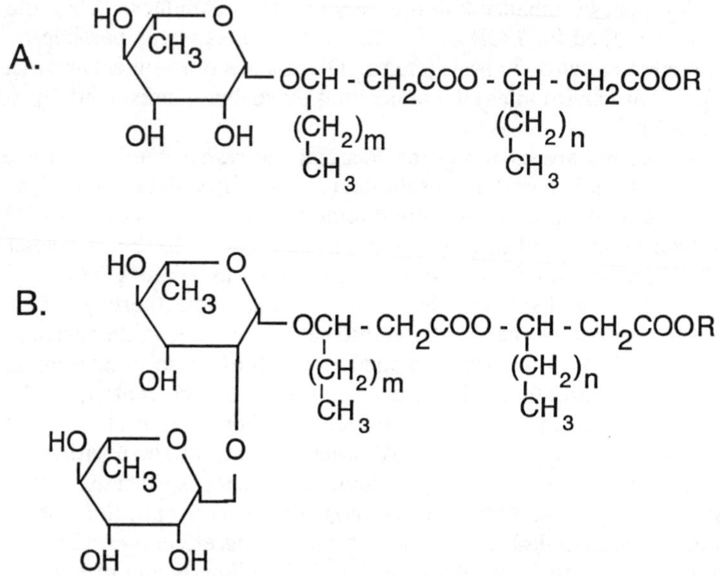

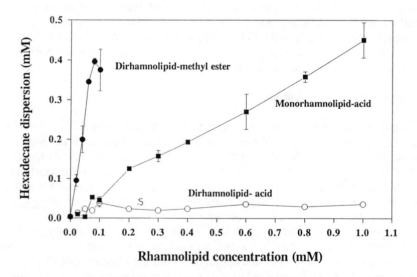

Figure 1.   The basic rhamnolipid structures A) monorhamnolipid where
             R = H or CH₃, and B) dirhamnolipid where R = H or CH₃.

Figure 2.   Solubilization (dispersion) of hexadecane by
             monorhamnolipid-acid, dirhamnolipid acid, and
             dirhamnolipid-methyl ester.

12), monorhamnolipid acid (1 mM, 500 mg l^{-1})) removed approximately 22% of residual hexadecane from sand columns in 120 pore volumes. A comparison of monorhamnolipid with two synthetic surfactants reveals that sodium dodecyl sulfate was ineffective in removing residual, and that polyoxyethylene(20) sorbitan monooleate was 4-fold less effective than the monorhamnolipid (see Figure 3). Removal of the hexadecane residual was primarily by mobilization.

Although biosurfactants are effective solubilization and emulsification agents, the more intriguing aspect of biosurfactants is the potential for enhanced biodegradation of organic compounds during remediation. We have shown that rhamnolipids are effective in increasing biodegradation rates of organic compounds in batch solution culture (9,10,13). There are two important factors in determining enhancement of biodegradation; the effect of the rhamnolipid on solubilization/emulsification (or bioavailability) of substrate, and the effect of the rhamnolipid on cell surfaces and the uptake of solubilized substrate. Enhancement of biodegradation, like enhancement of solubilization/emulsification, is dependent on the structure of the rhamnolipid. We have observed increases in hexadecane biodegradation of up to 30-fold by the dirhamnolipid methyl ester (0.05 mM), and 22-fold by the dirhamnolipid acid (0.05 mM).

For all organisms and organic compounds tested thus far, the dirhamnolipid methyl ester has been most effective in solubilization/emulsification and enhancement of biodegradation. One problem in working with the dirhamnolipid methyl ester is its low water solubility (< 0.04 mM). We have found that the solubility of the dirhamnolipid methyl ester can be increased by mixing it with dirhamnolipid acid. The amount of organic solubilized/emulsified by such a mixture lies between those amounts obtained by each rhamnolipid separately. Therefore, we speculate that one reason bacteria produce mixtures of surfactants is to optimize the properties of each. We are continuing to evaluate biosurfactants for use in remediation, focusing on sorptive interactions of rhamnolipid with soil and on in situ biosurfactant production.

Biosurfactants: Cyclodextrins

Cyclodextrins are cyclic oligosaccharides formed from the degradation of starch by bacteria. They have a hydrophilic outer shell and a hydrophobic interior cavity. This characteristic provides cyclodextrins with an excellent capacity to solubilize organic compounds. For example, Wang and Brusseau (14) showed that a 0.7% solution of hydroxypropyl-β-cyclodextrin (HPCD) increased the apparent aqueous solubilities of several organic compounds by factors from approximately 10 to 1000. This is illustrated in Figure 4a, where data for anthracene are reported. The solubilization power of HPCD was found to be much greater than that of miscible cosolvents (e.g., ethanol) and to be somewhat less than that of typical synthetic surfactants (3). With this level of solubilization power, cyclodextrins have potential to reduce sorption and enhance transport of highly sorptive organic contaminants. This ability is

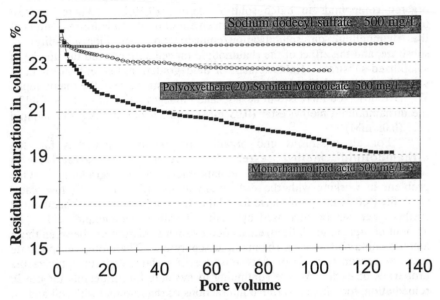

Figure 3.   Displacement of residual hexadecane by surfactant solutions. In this experiment, a column containing 40-50 mesh Accusand was saturated with hexadecane and then flushed with water to form a residual hexadecane saturation of 22%. The column was then flushed with a surfactant solution to remove residual hexadecane.

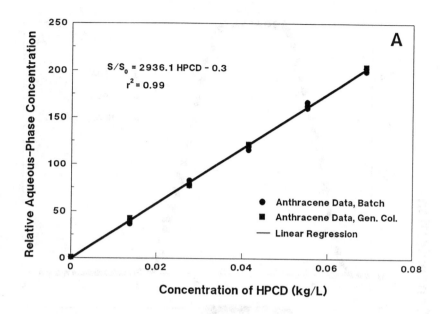

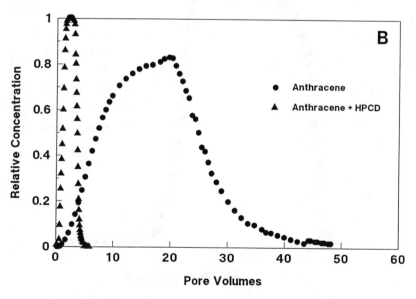

Figure 4.   (A) Solubilization of anthracene by a solution containing hydroxypropyl-β-cyclodextrin (HPCD); (B) The influence of HPCD on transport of anthracene through a column packed with a sandy soil. Adapted from Wang and Brusseau (1993) and Brusseau et al., (1994).

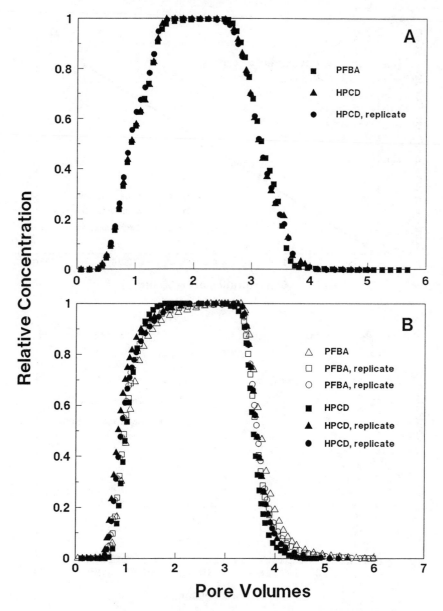

Figure 5.   Transport of hydroxypropyl-β-cyclodextrin (HPCD) and
               pentafluorobenzoate (PFBA) through (A) a column packed
               with a sandy soil; (B) a column packed with a high organic
               carbon content soil. Adapted from Brusseau et al., 1994.

illustrated in Figure 4b, which shows breakthrough curves for anthracene transport with and without HPCD.

We refer to cyclodextrin as a biosurfactant because they are produced by microbial action and they cause a reduction in surface tension. However, it must be noted that cyclodextrins do not form micelles, which may be advantageous in some situations. Cyclodextrins have several additional properties that make them prime candidates for use in chemically enhanced remediation. First, it is unlikely that they will experience significant pore exclusion phenomena during transport in porous media given their relatively small size ($^-$1.5 nm outer diameter). Data supporting this was reported by Brusseau et al. (15) and is reproduced in Figure 5. Breakthrough curves measured for the transport of HPCD through two soils were identical to breakthrough curves measured for a nonreactive tracer (pentafluorobenzoate). This shows the absence of retardation and of pore exclusion. Thus, they appear to be nonreactive with soil, which is a critical factor for the effectiveness of chemical additives. Third, they are not toxic to humans or to bacteria, they are resistant to hydrolysis, and are produced at commercial quantities. Considering all of these factors, cyclodextrins deserve continued investigation for use in chemically enhanced remediation.

Cosolvents

The ability of miscible solvents to increase the apparent aqueous solubility and to reduce the sorption of low-solubility organic compounds has been widely demonstrated (see 3 and 16 for recent reviews). These properties serve as a basis for the possible use of miscible solvents to enhance removal of organic contaminants by in situ flushing (17-19). The enhanced desorption and transport induced by solvents is illustrated in Figure 6, which shows elution curves for removal of anthracene from a sandy soil in the presence of varying amounts of methanol. Retardation decreases as the methanol fraction increases, with almost no retardation at a methanol fraction of 0.7.

The sorption of miscible organic liquids by soil is generally extremely low. Little sorption is expected for compounds such as methanol and ethanol because of their polarity and large (infinite) aqueous solubility. The minimal sorption of alcohols has been widely demonstrated in the chromatography literature. Limited data for soil systems has also shown negligible sorption of alcohols (cf., 17,20,21). Hence, these compounds will be minimally retarded and will travel through the subsurface at essentially the velocity of water. This large mobility can be a useful characteristic. For example, alcohols may be useful as an "early warning" sign of the impending arrival of a contaminant plume emanating from a fuel spill. In regard to the use of alcohols for in-situ soil washing, the greater mobility means that an injected pulse of alcohol may be able to overtake a plume of a retarded solute.

Alcohols such as methanol have been reported to be biodegradable under both aerobic and anaerobic conditions (cf., 22-24). However, the concentrations of alcohol at which biodegradation occurred were less than 1%.

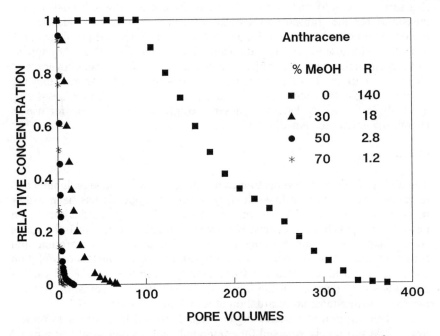

Figure 6.   The influence of methanol on elution of anthracene from a column packed with a sandy soil. Data from Brusseau et al., 1991.

Large concentrations (> 3-10%) of alcohol are generally considered to be toxic to most microorganisms and therefore not biodegradable. Thus, it would be desirable to recover residual alcohol from the subsurface when operating a remediation system.

Comparison of Additives

The primary criterion by which chemical-enhancement additives are normally judged is their solubilization potential. A comparison on a mass basis of the relative degree to which the aqueous-phase concentration of organic contaminants is enhanced by the various additives discussed in this paper favors biosurfactants, which perform similarly to synthetic surfactants (7). The solubilization power of cyclodextrins is somewhat less than that of surfactants. The solubilization power of miscible cosolvents is much less than that of the other two. However, a comparison of this type can be misleading without considering such factors as potential interactions between the additive and the porous media. It is well known, for example, that surfactant molecules can sorb to surfaces of solids (cf., 25-27), thereby reducing the concentration of additive available for solubilizing the contaminant. In addition, surfactants may precipitate under certain conditions. In contrast, most subsurface solids have a low affinity for cyclodextrins and for miscible solvents such as ethanol. Thus, it is possible that, whereas the "active" mass of a surfactant may be significantly less than the total mass injected into the subsurface, that of a cyclodextrin or miscible solvent may be essentially the same.

The interaction of the additive with microorganisms in the environment must be considered as well. As previously discussed, the additive can increase the bioavailability of contaminants, resulting in enhanced biodegradation and an overall reduction in contaminant concentration. There is evidence that the choice of additive is very important if stimulation of biodegradation is a remediation goal. This is because there seem to be specific interactions between additives and microorganisms that in some cases, e.g. rhamnolipid, lead to increased biodegradation. However, additives can also inhibit microbial activity due to toxic effects of the additive on microorganisms or by interfering with uptake of contaminant from contaminant-additive complexes (3,7,11). For example, large concentrations of cosolvents may kill microbial populations as previously discussed. It is also important to consider the biodegradability of the additive itself in order to evaluate the lifetime of the additive in the subsurface. Significant degradation of an additive can have several consequences. These include depletion of the effective concentration of the additive, and an increase in biomass that can cause plugging of pores resulting in changes in the movement of water and organic contaminants (cf., 28).

One aspect of the use of biosurfactants in subsurface remediation which has not yet been investigated but deserves mention, is in situ biosurfactant production. In situ biosurfactant production may potentially be stimulated in any environment colonized by microorganisms. This approach could allow very site specific production of biosurfactants which may be very useful in remediation of certain environments, for example contamination in zones of low hydraulic conductivity.

The impact of interactions between the additive, the solid phase, and microorganisms are important factors to consider, as discussed above. However, there are several other factors that should also be considered when selecting an enhancement agent. In this regard, cosolvents have benefits that surfactants and cyclodextrins may not have.

First, the addition of a cosolvent increases the magnitude of the desorption rate coefficient (not to be confused with an increase in the rate of desorption), thereby reducing the time required to attain equilibrium (17). This reduction in the degree of nonequilibrium would result in reduced tailing during pumping. This, in turn, would decrease the volume of water and the time required to remove the contaminant by flushing. Rate-limited desorption may impose a significant constraint on the efficacy of pump-and-treat remediation in certain cases. If so, the ability of a cosolvent to reduce the degree of nonequilibrium would be a major attribute.

Second, cosolvents may be able to "extract" the highly retained, aged contaminants that have been observed in field studies. This hypothesis is based on the results of solvent extraction techniques used in the analysis of contaminated soils (cf., 29) and on the results of experiments that evaluated the effect of cosolvents on the desorption of organic compounds (17,30,31).

In addition, both cyclodextrins and cosolvents may be able to access contaminant that is residing in low hydraulic-conductivity domains such as clay lenses. During a pump-and-treat remediation, contaminants in these domains are removed partly through diffusion. The clay particles provide a large surface area with which a surfactant may interact and thereby reduce its availability for enhancing contaminant removal. In addition, the sorption of the surfactant can enhance the retention of the organic solutes by providing an increase in stationary-phase organic carbon. Surfactant aggregates may possibly be excluded from the smaller pore-size domains, which would limit accessibility. Cyclodextrins and miscible cosolvents generally do not sorb to solid surfaces and, because of their small size, would not be excluded from any pore domains in which contaminants would be found. Thus, in comparison to surfactants, cyclodextrins and cosolvents may have a greater potential for enhancing the release of contaminants trapped in fine-grained media.

Conclusion

The selection of which additive to use for a subsurface-remediation project is dependent on properties of the site, of the target contaminant, and the cleanup objectives. For removal of immiscible organic liquids, surfactants have a distinct advantage, in comparison to cosolvents and cyclodextrin, in that surfactants can induce mobilization of immiscible liquids through a reduction in interfacial tension. This latter property may be an advantage or disadvantage depending on the system of interest. Since mobilization is generally much more rapid at removing immiscible liquids, it would be the preferred approach. However, if possible escape of the mobilized contaminant is of concern (e.g., mobilization of a dense liquid in a system with no confining layer below the zone of contamination), the application of a surfactant may be limited to the solubilization mode. When selecting an additive for use in a solubilization-based approach (e.g., for removal of highly sorbed solute or when emulsification is undesirable), the performance evaluation must consider factors other than solubilization power. It is in these factors where cyclodextrins and cosolvents have advantages in comparison to surfactants.

Considering the preceding discussion, it is clear that each of the additives have associated advantages and disadvantages. Which additive may be best suited for a specific application must be evaluated for that particular system. Finally, it must be kept in mind that the use of surfactants and cosolvents to enhance flushing will generally be limited primarily to smaller scale problems such as source control/removal because of economic constraints.

Acknowledgements

This research was supported by a grant from the U.S. Department of Energy Subsurface Science Program.

References

1. Keely, J.F., **Performance Evaluations of Pump-and-Treat Remediations**, Ground Water Issue Paper, U.S. Environ. Prot. Agency, Wash. D.C., Oct. 1989.
2. MacKay, D.M.; Cherry, J.A. Environ. Sci. Technol. 1989, 23, 630.
3. Brusseau, M.L.; **Complex Mixtures and Groundwater Quality**, Environmental Research Brief, U.S. Environ. Prot. Agency, Washington, D.C., 1993.
4. Palmer, C.D.; Fish, W. **Chemical enhancements to pump-and-treat remediation**, Groundwater Issue Paper, U.S. Environ. Prot. Agency, Wash. D.C., 1992.
5. Sabatini, D.A.; Knox, R.C., eds., **Transport and Remediation of Subsurface Contaminants**, ACS Symposium Series 491, American Chemical Society, Wash., D.C., 1992.
6. McCarthy, J.F.; Wobber, F.J., eds. **Manipulation of Groundwater Colloids** for Environmental Restoration, Lewis Publ., Ann Arbor, MI, 1993.
7. Miller, R.M. in: **Bioremediation - Science & Applications**, Skipper, H., ed., Soil Science Society of America special publication, Madison, WI, 1994 in press.
8. Zajic J.E.; Panchel, C.J. CRC Critical Reviews in Microbiology, 1976, 5, 39.
9. Zhang, Y.; Miller, R.M., Appl. Environ. Microbiol., 1992, 58, 3276.
10. Zhang, Y.; Miller, R.M., Appl. Environ. Microbiol., 1994a, 60, 2101.

11.   Miller, R.M.; Jimenez, I.; Bartha, R. in: **Nonmedical Applications of Liposomes**, Barenholz, Y.; Lasic, D.D., eds., CRC Press, Boca Raton, 1994 in review.

12.   Bai, G-Y.; Brusseau, M.L.; Miller, R.M. , Environ. Sci. Technol., 1994 in review.

13.   Zhang, Y.; Miller, R.M., Appl. Environ. Microbiol., 1994b in review.

14.   Wang, X.; Brusseau, M.L. Environ. Sci. Technol. 1993, 27, 2821.

15.   Brusseau, M.L.; Wang, X.; Hu, Q. Environ. Sci. Technol. 1994, 28, 953.

16.   Rao, P.S.C.; Lee, L.S.; Wood, A.L. **Solubility, Sorption and Transport of Hydrophobic Organic Chemicals in Complex Mixtures**, Environmental Research Brief, U.S. Environ. Prot. Agency, Washington, D.C., 1991.

17.   Brusseau, M.L.; Wood, A.L.; Rao, P.S.C. Environ. Sci. Technol. 1991, 25, 903.

18.   Rixey, W.G.; Johnson, P.C.; Deeley, G.M.; Byers, D.L.; Dortch, I.J. pp. 387-409 in: **Hydrocarbon Contaminated Soils**, Calabrese, E.J.; Kostecki, P.T., eds., Lewis Publ., Ann Arbor, MI, 1992.

19.   Augustijn, D.; Jessup, R.E.; Rao, P.S.C.; Wood, A.L. J. Environ. Engin. 1994, 120, 42.

20.   Garrett, P.; Moreau, M.; Lowry, J.D. pp. 227-238 in: Petroleum Hydrocarbons and Organic Chemicals in Ground Water: Prevention, Detection, and Restoration, Nat. Water Well Assoc., Dublin, OH, 1986.

21.   Wood, A.L.; Bouchard, D.C.; Brusseau, M.L.; Rao, P.S.C. Chemosphere. 1990, 21, 575.

22.   Colby, J.; Dalton, H.; Whittenburg, R. Ann. Rev. Microbiol. 1979, 33, 481.

23.   Lettinga, G.; deZeeuw, W.; Ouborg, E. Water Res. 1981, 15, 171.

24.   Novak, J.T.; Goldsmith, C.D.; Benoit, R.E.; O'Brien, J.H. pp. 71-85 in: Degradation, Retention, and Dispersion of Pollutants in Ground Water, Inter. Assoc. Water Pollution Control Research, Great Britain, 1985.

25.   Ducreux, J.; Bocard, C.; Muntzer, P.; Razakarisoa, O.; Zilliox, L. Water Sci. Tech. 1990, 22, 27.

26.   Kan, A.T.; Tomson, M.B. Environ. Toxic. Chem. 1990, 9, 253.

27.   Jafvert, C.T.; Heath, J.K. Environ. Sci. Technol. 1991, 25, 1031.

28.   Vandevivere, P.; P. Baveye. Soil Sci. Soc. Amer. J. 1992, 56, 1.

29.   Sawhney, B.L.; Pignatello, J.J.; Steinberg, S.M. J. Environ. Qual. 1988, 17, 149.

30.   Freeman, D.H.; Cheung, L.W. Science. 1981, 214, 790.

31.   Nkedi-Kizza, P.; Brusseau, M.L.; Rao, P.S.C.; Hornsby, A.G. Environ. Sci. Technol. 1989, 23, 814.

RECEIVED December 13, 1994

Chapter 8

Fate of Linear Alkylbenzene Sulfonate in Groundwater

Implications for In Situ Surfactant-Enhanced Remediation

Larry B. Barber, II[1], Carolyn Krueger[2], David W. Metge[1],
Ron W. Harvey[1], and Jennifer A. Field[3]

[1]U.S. Geological Survey, 3215 Marine Street, Boulder, CO 80303
[2]Department of Agriculture Chemistry and [3]Department of Chemistry,
Oregon State University, Corvallis, OR 97331

A small-scale natural-gradient tracer test was conducted to determine the transport behavior of linear alkylbenzene sulfonate (LAS) surfactants in oxygen-depleted groundwater. LAS transport was similar to that of the conservative tracer bromide and had a retardation factor of about 1.1. During the 45-day experiment, LAS was not significantly biodegraded. Sorption to the aquifer sediments changed the LAS mixture composition during transport due to increasing retardation of the long-chain homologs. The change in homolog composition may alter the effectiveness of LAS to facilitate organic contaminant transport. The abundance of free-living bacteria increased by a factor of 3 in the presence of the injected LAS indicating that surfactants can alter the subsurface microbial populations. The relatively unretarded transport and persistence of LAS in the oxygen-limited aquifer are favorable characteristics for subsurface remediation.

A variety of surfactants have been evaluated as agents to enhance remediation of aquifers contaminated by hydrophobic-organic compounds (*1-8*). Although non-ionic surfactants have been most widely investigated, anionic surfactants also are candidates. The advantage of anionic surfactants is their low potential for sorption onto aquifer sediments due to repulsion between the anionic-head group and negatively-charged sediments (*9*). While the solubility enhancement characteristics of most surfactants proposed for use in aquifer cleanup have been determined in the laboratory, few studies report on the transport and behavior of surfactants under field conditions. Laboratory data do not always agree with results obtained from field studies, and the lack of field data is a limitation in assessing the viability of the *in situ* approach. Field studies provide "ground truth" for evaluating hypotheses and mathematical models. Small-scale tracer tests that focus

0097–6156/95/0594–0095$12.00/0
© 1995 American Chemical Society

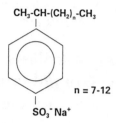

Figure 1. Chemical structure of LAS showing the 2-phenyl isomer.

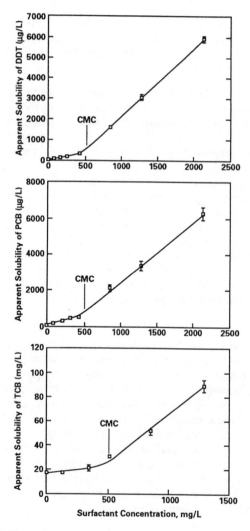

Figure 2. Solubility enhancement of DDT, PCB, and trichlorobenzene by LAS (from reference 13). CMC= Critical micelle concentration.

on the behavior of surfactants are needed before implementation of pilot and full-scale surfactant-enhanced remediation schemes.

Linear-alkylbenzene sulfonates (LAS) are the highest volume anionic surfactants used in domestic and commercial detergent formulations (*10*). LAS consists of a series of homologs ranging from 10-14 carbon atoms in the aliphatic chain, and each homolog has a series of phenyl-position isomers (Figure 1). The 26 different components in commercial LAS mixtures have slightly different physicochemical and biodegradation characteristics. The octanol/water partition coefficients (K_{ow}) for several LAS homologs are given in Table I. Commercial LAS mixtures also contain synthesis byproducts such as anionic dialkyltetralin and indane sulfonates, and neutral oils such as bis(alkylphenyl)-sulfones and unreacted linear alkylbenzenes (*11,12*).

Several features of LAS make it suitable for surfactant-enhanced *in situ* remediation. First, LAS has been shown to enhance the solubility of hydrophobic organic compounds (*1,13*). Second, LAS has a high water solubility (~200 mg/L) and is poorly sorbed to aquifer sediments, which can result in rapid subsurface transport (*14*). Third, LAS is relatively nontoxic to humans and aquatic organisms (*15*). Lastly, LAS is an attractive candidate because of its availability and relatively low cost. However, there are limitations to the use of LAS, such as the formation of viscous emulsions that can decrease permeability in an aquifer (*1*) and susceptibility to precipitation in the presence of inorganic cations such as Ca^{+2} (*9*).

Unlike many other surfactants, an extensive data base exists on the fate of LAS in the environment (*14,16-18*). It is well documented that LAS is readily biodegradable under aerobic conditions, such as those that exist in many surface waters, where the biodegradation half-lives range from a few hours to several days (*18-22*). However, LAS is persistent under the low-oxygen conditions that exist in anaerobic sewage-sludge digestors (*23*) and in sewage-contaminated groundwater (*16,24-27*).

Solubility enhancement of hydrophobic-organic compounds by LAS was investigated by Chiou et al. (*13*) who reported that, in the presence of LAS above the critical micelle concentration (CMC ~500 mg/L), the solubility of DDT and PCB is increased by several orders of magnitude and the solubility of trichlorobenzene is affected to a lesser extent (Figure 2). At concentrations below the CMC, LAS with 1-2% neutral-oil content increased solubility by a factor of 2, whereas oil-free LAS (CMC ~390) showed no solubility enhancement. The presence of neutral oil also has a significant effect on the viscosity and solubility of LAS solutions (*11,12*). Bolsman et al. (*28*) reported that alkyl-chain length, point of attachment of the phenyl group to the alkyl chain, and aromatic-substitution position of LAS can vary the rate of hydrophobic-organic contaminant solubilization by an order of magnitude.

Sorption of LAS to sediments is variable as indicated by the range of LAS sediment/water distribution coefficients (K_d) and sediment organic carbon normalized distribution coefficients (K_{oc}) given in Table I. This wide range of values reflects the dependency of LAS sorption on sediment properties such as soil-organic carbon, and the different sorption characteristics of the various homologs

Table I. Physicochemical, Sorption, and Transport Characteristics of LAS

Homolog	$\log K_{ow}$[a]	$\log K_{oc}$	
C_{10} LAS	1.23	0.57[b]- 1.38[c]	
C_{11} LAS	1.26	0.60 - 1.40	
C_{12} LAS	1.96	1.23 - 1.90	
C_{13} LAS	2.52	1.74 - 2.30	
C_{14} LAS	2.73	1.93 - 2.46	
Reference	K_d (L/kg)[d]	$\log K_{oc}$	R_f[e]
Hand and Williams[f]	3 - >10,000	3.65 - 6.75	14 - >10,000
Hand et al.[g]	11 - 5,900	3.62 - 6.17	49 - >10,000
Di Toro et al.[h]	3 - 26,000	2.96 - 4.81	1.01 - 113
Schwarzenbach and Westall[i]	0.006 - 0.057	1.38 - 2.46	1.03 - 1.25
Chiou et al.[j]	0.001 - 0.013	0.57 - 1.93	1.00 - 1.06
Results from this study[k]	0.027	1.43 - 2.43	1.11 - 1.12

[a]. Octanol/water partition coefficient (K_{ow}) data from ref. 29; [b]. sediment organic carbon/water distribution coefficients (K_{oc}) calculated from equation in ref. 32; [c]. K_{oc} calculated from equation in ref. 33; [d]. sediment/water distribution coefficient, $K_d = K_{oc} \cdot f_{oc}$, where f_{oc} = fraction organic carbon; [e]. relative retardation factor, $R_f = 1 + (\rho \cdot K_d)/\theta$, ρ = bulk density = 1.64 g/cm^3, θ = porosity = 0.38; [f]. ref. 29, lab measurements, SOC = 0.9-3.5, C_{10}-C_{14} homologs and isomers; [g]. ref. 30, field measurements, SOC = 1.6-3.9, C_{10}-C_{14} homologs and isomers; [h]. ref. 31, lab measurements, SOC = 0.34-40, C_{10}-C_{14} homologs and isomers; [i]. calculated from equation in ref. 32, f_{oc} = 0.0001-0.001; [j]. calculated from equation in ref. 33, f_{oc} = 0.0001-0.001; [k]. K_{oc} calculated using f_{oc} = 0.0001-0.001.

and isomers (29-33). In sediments containing little organic carbon, LAS is poorly sorbed and readily transported.

In this report, we present the results from a small-scale natural-gradient tracer test conducted in a low-oxygen (<0.1 mg/L) zone of sewage-contaminated groundwater. The test was designed to determine the transport characteristics of a LAS mixture and to determine if LAS biodegradation occurs under the *in situ* conditions of the aquifer. A low-oxygen zone was selected (for this first in a series of tracer tests) because minimal biodegradation was expected to occur, thus focusing the test on LAS transport. The approach of using natural-gradient *in-situ* tracer experiments is an important link in the extension of laboratory measurements of biodegradation and transport to field scale processes, because the experimental conditions are defined by the actual hydrological, geochemical, and microbiological conditions of the aquifer.

Description of the Field Site

The field study was conducted in a sand-and-gravel glacial-outwash aquifer located on Cape Cod, near Falmouth, Massachusetts (Figure 3). This site has an extensively studied sewage plume (*34-40*). A number of investigations have specifically assessed the fate of LAS and related compounds in the plume (*16,24-27*). Small- and intermediate-scale natural-gradient tracer experiments have been conducted to examine the fate of inorganic and microbiological constituents (*41-47*). These tracer experiments used a three-dimensional network of multilevel wells that encompass a wide range of geochemical and hydrological conditions (*48*).

Field Experiment

The tracer experiment was conducted between June 6 and July 20, 1993, at site F347, located approximately 300 m downgradient from the sewage-disposal beds (Figure 3). The test site has 9 multilevel wells (Figure 4), each consisting of 15 sampling ports spaced at 0.5-m intervals between 4.5 and 12.6 m above mean sea level (MSL). The land-surface altitude is approximately 18.2 m above MSL, and the water table was located about 3.9 m below land surface. The hydraulic gradient was about 0.002, and the mean direction of flow was 163° east of north. The sampling ports of the wells span a distance of about 8 m, and include the overlying uncontaminated groundwater and a portion of the sewage plume. Dissolved oxygen ranged from <0.1-10 mg/L and specific conductance ranged from 50-250 µS/cm over this interval (Figure 5).

The tracer test involved collecting contaminated groundwater from a zone 7.4 m above MSL (6.9 m below the water table), amending the water with LAS and bromide (used as a conservative tracer), and reinjecting the tracer mixture into the same sampling port from which it was collected. The injection was conducted over a 3-hour period. Special precautions were taken to maintain the groundwater used for the injectate under *in situ* conditions. Approximately 187 L of groundwater was removed from the sewage plume using a peristaltic pump equipped with Norprene tubing. The background concentration of dissolved oxygen was <0.1 mg/L, specific conductance was 250 µS/cm, and LAS was <0.01 mg/L. The groundwater was collected in a gas-impermeable bag (Aerotech, Ramsey, NJ) and placed in a water filled pit to maintain the injectate at ambient aquifer temperature (12-14° C). Prior to adding the groundwater, the bag was purged with nitrogen and a solution of LAS and sodium bromide in oxygen-free distilled water was added. The final groundwater injectate solution contained 12 mg/L LAS and 100 mg/L bromide. The LAS was provided by Vista Chemicals (Austin, TX) and had the following homolog distribution: 22% C_{10}, 39% C_{11}, 29% C_{12}, 9% C_{13}, and 1% C_{14}.

Wells M11 and M10 located 4.5 m downgradient from the injection well and well M5 located 6.9 m downgradient from the injection well (Figure 4) were monitored daily for 45 days to evaluate breakthrough of LAS and bromide. Samples were collected by peristaltic pump. Dissolved oxygen was measured

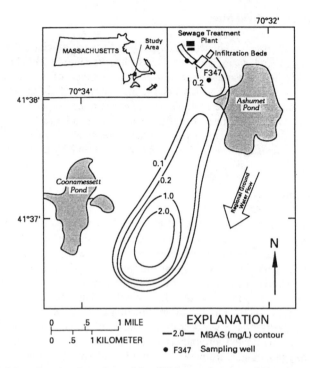

Figure 3. Map showing location of the field experiment and the sewage plume (from reference 16) as defined by methylene blue active substances.

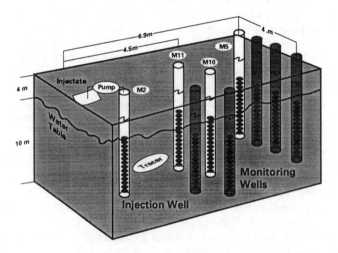

Figure 4. Diagram showing configuration of the multilevel well array used for the tracer test. Vertical and horizontal distances are not to scale. Bromide and LAS were detected in the wells shown in white.

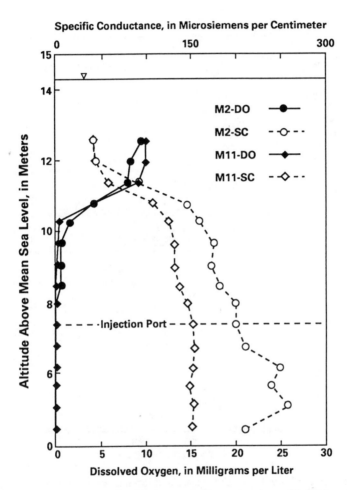

Figure 5. Geochemical conditions in the aquifer at the time of injection. Data are from well M2, where groundwater was withdrawn and reinjected, and well M11 which is 4.5 m downgradient from the injection well. Profiles for dissolved oxygen (DO) and specific conductance (SC) are shown.

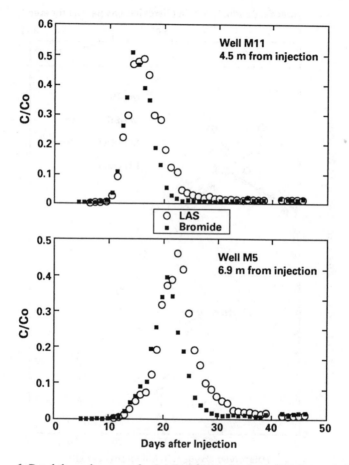

Figure 6. Breakthrough curves for LAS and bromide in wells M11 and M5 for the *in situ* natural-gradient tracer experiment conducted in June 1993. Concentrations measured daily (C) are normalized against the injectate concentration (C₀).

directly in the pump discharge by using a colorimetric ampule assay system (CHEMetrics, Calverton, VA). Samples for bromide analysis were collected in 60-mL polyethylene bottles, samples for LAS analysis were collected in 500-mL acid-washed polyethylene bottles and preserved with 1% formalin, and samples for bacteria enumeration were collected in 60-mL polyethylene bottles and preserved with 3% formalin. Bromide analysis was performed on site using an ion-selective electrode after the sample temperature had equilibrated to ~25° C. Measurement of LAS was performed on site by the methylene blue active substance method which involved adjusting the pH of a 5-mL sample to <2, adding methylene blue to form an ion pair with LAS, extracting the ion pair into chloroform, and measuring the adsorption of the chloroform extract at 635 nm. Specific analysis for LAS was performed at the Oregon State University laboratory using the ion-pair/injection-port derivatization method of Field et al. (49). LAS was isolated from the groundwater by passing 5-50 mL of sample through a C_{18}-modified Empor disk, extracting the disk with an ion-pair reagent consisting of 0.5 M tetrabutylammonium hydrogen sulfate in chloroform, and analyzing the extracts by gas chromatography with flame-ionization detection. Concentrations of LAS were calculated from integrated peak areas using 1-phenyl C_8-LAS (Aldrich, Milwaukee, WI) which was added as a surrogate standard prior to extraction. Bacteria enumeration was performed by staining the bacteria with a 5 μg/L solution of 4'6-diamidino-2-phenylindole (DAPI) for 1 hour, filtering through a 0.22-μm membrane, and directly counting the individual bacteria using epifluorescent microscopy (43).

Results

Concentration histories (breakthrough curves) for LAS and bromide at the wells downgradient from the injection were constructed by normalizing the concentrations measured daily (C) to that of the injectate (C_0). Breakthrough curves for the 4.5 m and 6.9 m wells were nearly Gaussian in shape for both bromide and LAS (Figure 6). Peak LAS concentrations were 5.7 mg/L and 5.6 mg/L at the 4.5 m and 6.9 m wells, which corresponded to 48% and 47% of the injectate LAS concentrations. Peak bromide concentrations were 50 mg/L and 40 mg/L at the 4.5 m and 6.9 m wells, which corresponded to 50% and 40% of the injectate bromide concentrations.

The method of moments (50,51) was used to calculate the mass in solution and the average travel time for the center of mass of bromide, the conservative tracer, and LAS. The zeroth moment (M_0), or the integral area under the bromide and LAS breakthrough curves, was calculated by summing the C/C_0 values. M_0 provides an estimate of the mass for each solute in the system. The integral area for bromide in the 4.5 m well was 3.0 and the area for LAS was 3.7 (Table II). The integral area for bromide in the 6.9 m well was 2.6 and the area for LAS was 3.1. The ratios of the integral areas of LAS to bromide were 1.3 for the 4.5 m well and 1.2 for the 6.9 m well. These values indicate that no significant removal of LAS occurred under the conditions of the tracer test. Values less than 1 would suggest a removal of LAS mass relative to bromide.

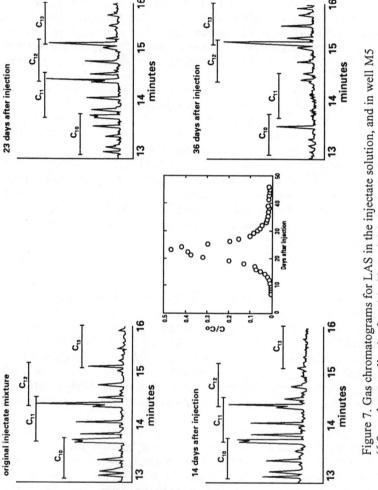

Figure 7. Gas chromatograms for LAS in the injectate solution, and in well M5 (6.9 m downgradient) after 14, 23, and 36 days of transport in the aquifer. Note that the large peak in the C_{10}-LAS window for the day 36 chromatogram is not present in the original LAS mixture.

Table II. Values for the Zeroth Moment (M_0) and First Moment (M_1) for Bromide and LAS in Observation Wells 4.5 Meters (M11) and 6.9 Meters (M5) Downgradient from the Injection Well

	Well M11		Well M5	
	Bromide	LAS	Bromide	LAS
Integral area (M_0)	2.96	3.69	2.58	3.05
Average travel time (days) of center of mass (M_1)	16.15	17.98	20.81	23.25
Retardation factor	1[a]	1.11	1[a]	1.12

[a] By definition, the retardation factor of the conservative tracer is equal to 1.

The first moments (M_1) for the bromide and LAS breakthrough curves were calculated to locate the day corresponding to the centers of mass. To estimate the average travel time for the center of mass for each solute, the distance from the injection well to the observation well was divided by the days after injection that corresponded to the center of mass. Using the travel time for the center of bromide mass to reach the 6.9 m well, we calculate an average groundwater flow velocity of 0.33 m/day. For both wells, the travel time for bromide was less than that of LAS. The ratio of the travel time for the center of LAS mass to that of bromide provides an estimate of the retardation factor for LAS under field conditions. A retardation factor of 1.1 was calculated for both the 4.5 m and the 6.9 m wells (Table II). The average K_d for LAS calculated from the retardation factors determined for the two wells was 0.027 L/kg (Table I). The observed K_d, K_{oc}, and retardation factor (R_f) values are consistent with those calculated from empirical equations relating sorption to the K_{ow} of the compound and the sediment organic carbon content, and are considerably less than those measured for surface-water sediments (Table I). The higher K_d values for surface sediments are attributed to their higher organic carbon content (typically greater than 0.5%), whereas the aquifer sediments at the Cape Cod site have organic-carbon contents less than 0.1% (*52*). Sorption to the aquifer sediments is less than that reported for surface sediments, even when normalized to the sediment organic-carbon content.

The retardation and apparent tailing of LAS relative to bromide (Figure 6) is attributed to sorption to the sediments. The observed tailing for total LAS is the result of chromatographic separation of the LAS mixture components (homologs and isomers) during transport in the aquifer (Figure 7). In the initial part of the LAS breakthrough curve (day 14) for the 6.9 m well, the LAS mixture was enriched in C_{10} homologs relative to that of the injectate and no C_{13} homologs were detected. In the middle portion of the curve (day 23) all homologs were present, although the distributions were skewed towards the lower homologs compared to the original injectate. In the tail portion of the curve (day 36) the mixture was enriched in C_{12} and C_{13} homologs. Note that the peak in the C_{10}-LAS window at about 13.5 minutes for the day 36 chromatogram is an artifact and does not occur in the original LAS mixture. The separation of LAS homologs is

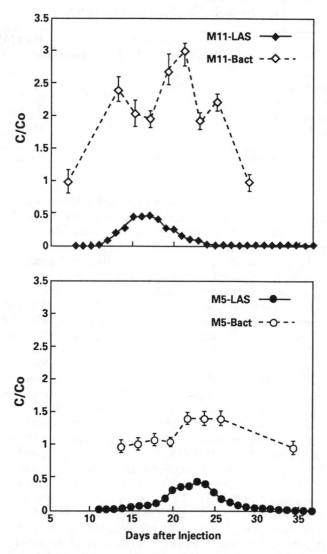

Figure 8. Relative LAS concentrations and abundance of free-living bacteria observed in wells M11 (4.5 m downgradient) and M5 (6.9 m downgradient). The bacteria abundance in the groundwater prior to the LAS breakthrough curve was taken to be C_0. Error bars represent one standard error based on duplicate analysis.

consistent with a hydrophobic mechanism where sorption increases with increasing chain length and increasing K_{ow} (29). The K_{oc} (calculated from reference 32) for C_{10}-LAS is ~10-times less than for C_{13}-LAS and results in a difference between the calculated R_f (assuming 0.1% organic carbon) of 1.10 versus 2.24. The predicted difference in R_f between the homologs is consistent with their observed separation in the aquifer. A similar phenomenon of chromatographic separation was observed at this site in an earlier experiment involving a mixed population of microspheres that differed in surface chemistry characteristics (43).

The indigenous bacteria in the aquifer were affected by the presence of LAS in the groundwater. At 4.5 m from the point of injection, LAS breakthrough was accompanied by a 3-fold increase (from 6.7×10^5 to $2.0 \times 10^6/$ mL) in abundance of unattached bacteria (Figure 8). The free-living bacteria concentrations were significantly different ($p = 0.01$, student t-test) from background concentrations across the entire breakthrough curve. In the 6.9 m well the effect was less evident, although the free-living bacteria concentrations were significantly higher ($p = 0.05$) than the initial concentrations 22-24 days after injection. The peak bacteria concentrations were slightly offset from the LAS and bromide breakthrough curves in both wells. The increase in bacteria abundance may be due to an increase in the growth rate of free-living bacteria, an increase in the detachment of adherent bacteria, or a combination of the two effects.

Although LAS is readily degraded in aerobic systems (19-21) it can have an inhibitory effect on bacterial growth at the high concentrations used in this test (15). It is possible that the addition of a labile substrate such as LAS can stimulate growth of indigenous bacteria, which would result in an increase in their abundance. The lack of readily-degradable organic carbon is a major limitation for bacterial growth in the contaminated and uncontaminated parts of the aquifer (53). Since bacteria under severe carbon limitation often grow faster at solid surfaces (54), adherent bacteria have an advantage over unattached bacteria in the aquifer. However, the advantage disappears with addition of a readily-degraded organic substrate. The partitioning of bacteria between aqueous and solid phases in aquifer sediments varies considerably in the presence of organic contaminants (55). There is a direct correlation between abundance of free-living bacteria and labile organic carbon within the sewage plume, and areas with the highest concentrations of labile organic carbon have the largest abundance of free-living bacteria (38). Therefore, any substantive addition (mg/L levels) of a readily-degraded organic substrate such as LAS could lead to concomitant increases in numbers of free-living bacteria. Harvey and George (56) report a bacteria generation time of 16 hours for the area where this experiment was conducted, indicating that bacteria growth potentially can occur within the time frame of the tracer test. Further field work is underway to delineate between bacterial growth and transport.

LAS can affect the abundance of free-living bacteria in the aquifer by promoting detachment of adherent populations from grain surfaces, and by preventing reattachment of bacteria present in the pore water. The results from this experiment indicate that LAS concentrations of only 1-2% of the CMC may affect the attachment and transport characteristics of indigenous subsurface bacteria. Preliminary results from static column experiments and other small-scale injection experiments indicate that LAS is moderately effective in causing detachment of fluorescently-labeled indigenous bacteria from the aquifer sediments (Harvey et al., unpublished). It is likely that LAS would be less effective in causing detachment of indigenous bacteria that had been attached for some period of time because of the large quantities of extracellular bridging polymers that can accumulate. Results from static column experiments indicate that LAS is much more effective in preventing bacterial deposition than it is in causing detachment once attachment to aquifer surfaces has taken place. The effect of LAS upon bacterial attachment behavior in the aquifer sediments is pH-dependent, and the greatest attachment occurs under acidic conditions (Harvey et al., unpublished). The pH of the groundwater at the depth of the injection was slightly acidic.

It also is possible that LAS has an inhibitory effect on protozoan populations that may otherwise prey upon the unattached bacteria. The population of protozoa in the aquifer sediments at the test site is quite large (up to 10^5/g) (57). However, very little is known about the effect of LAS upon groundwater protozoa.

Implications For *In Situ* Remediation

The results from this small-scale natural-gradient tracer test have several implications with respect to the use of anionic-surfactant mixtures such as LAS for *in situ* surfactant-enhanced aquifer remediation. Even though LAS is readily transported in the sand-and-gravel aquifer, sorption to the sediments can cause changes in the LAS mixture composition over time. Consequently, solubility enhancement determined in the laboratory may differ from that initially observed under field conditions. However, because LAS behaves similarly to the conservative tracer bromide, it can be readily flushed from the aquifer and thus is amenable to pump-and-treat operations.

The results of our field experiment indicate that LAS persists in low-oxygen groundwater. The slowness of LAS biodegradation under low-oxygen conditions can be viewed as either an asset or a detriment to a surfactant-enhanced aquifer-remediation scheme. From the practical standpoint, it is important that the surfactant not undergo rapid biodegradation relative to the time frame of the cleanup. For example, surfactants that biodegrade very quickly under anaerobic conditions could undergo primary degradation, lose their surfactant properties, and become unavailable to facilitate transport of the contaminant. On the other hand, low rates of LAS biodegradation under oxygen-limited conditions can be a drawback because biodegradation is the primary mechanism for LAS removal in the aquifer. However, once the cleanup is nearly complete, the aquifer might return to aerobic conditions, particularly if oxygen-enriched water is being pumped through the aquifer, and residual LAS could then biodegrade.

Because of its high mobility and slow biodegradation, LAS (as well as other surfactants) used for *in situ* remediation can become a groundwater contaminant if it escapes from the treatment site. When surfactant-based remediation is used in conjunction with a pump-and-treat system, it often is assumed that the injected surfactant and solubilized contaminants will be contained and completely recovered. However, under the complex conditions encountered under field situations and within the subsurface environment, it is likely that surfactant and contaminant recovery will be less than 100%. Consider a scenario where LAS is used in a pilot *in situ* cleanup similar to that reported by Abdul et al. (*6*). If LAS is added as a 1% (10,000 mg/L) solution (equivalent to 20 times its CMC) in 20,000 L of water, and 99% of the added mass is recovered, the remaining LAS could contaminate up to 4×10^6 L of groundwater at 0.5 mg/L (the foaming threshold concentration). This calculation does not consider hydrologic variables such as dispersion. Depending upon the conditions and duration of the remediation scheme, an extensive surfactant plume can be produced. At the study site, which has a groundwater-flow velocity of ~0.3 m/day and a LAS transport velocity similar to that of the groundwater, a 100 m long plume could result from a 1-year project. After remediation is halted, the LAS plume will continue to migrate. In addition to the potential for escape of the surfactant from the site is the more serious possibility of long-range contamination by the compounds that the remediation is attempting to removed from the aquifer. Compounds that initially have limited mobility can become highly mobile after surfactant treatment to facilitate their transport, and thus have a greater potential to contaminate the aquifer if not completely recovered.

Acknowledgments

We gratefully acknowledge the financial support of the National Geographic Society, Oregon State University Research Council, Vista Chemical, and Monsanto. We also acknowledge the field and logistical support provided by the Massachusetts District of the U.S. Geological Survey. Use of trade names is for identification purposes only and does not constitute endorsement by the U.S.G.S.

Literature Cited

1. American Petroleum Institute *API Publication 4390* **1985**, 59 p.
2. Ellis W.D.; Payne, J.R.; McNabb, G.D. *U.S. Environmental Protection Agency, EPA/600/2-85/129* **1985**, 84 p.
3. Nash, J.; Traver, R.P. *U.S. Environmental Protection Agency, EPA/600/9-86/022* **1986**, 208.
4. Abdul, A.S.; Gibson, T.L.; Rai, D.N. *Ground Water* **1990**, *28,* 920.
5. Abdul, A.S.; Gibson, T.L. *Environ. Sci. Technol.* **1991**, *25*, 665.
6. Abdul, A.S.; Gibson, T.L.; Ang, C.C.; Smith, J.C.; Sobczynski, R.E. *Ground Water* **1992**, *30*, 219.
7. Pennell, K.D.; Abriola, L.M.; Weber, W.J., Jr. *Environ. Sci. Technol.* **1993**, *27*, 2332.
8. Edwards, D.A.; Adeel, Z.; Luthy, R.G. *Environ. Sci. Technol.* **1994**, *28*, 1550.
9. Rouse, J.D.; Sabatini, D.A.; Harwell, J.H. *Environ. Sci. Technol.* **1993**, *27*, 2072.
10. Rapaport, R.A.; Eckhoff, W.S. *Environ. Toxicol. Chem.* **1990**, *9*, 1245.
11. Moreno, A.; Bravo, J.; Berna, J.L. *J. Am. Oil Chem. Soc.* **1988**, *65*, 1000.
12. Moreno, A.; Cohen, L; Berna, J.L. *Tenside Surfactants Detergents* **1988**, *25*, 216.

13. Chiou, C.T.; Kile, D.E.; Rutherford, D.W. *Environ. Sci. Technol.* **1991**, *25*, 660.

14. Holysh, M.; Paterson, S.; Mackay, D.; Bandurraga, M.M. *Chemosphere* **1986**, *15*, 3.

15. Kimerle, R.A. *Tenside Surfactants Detergents* **1989**, *26*, 169.

16. Field, J.A.; Barber, L.B., II; Thurman, E.M.; Moore, B.L.; Lawrence, D.L.; Peake, D.A. *Environ. Sci. Technol.* **1992**, *26*, 1140.

17. Games, L.M. In *Modeling the Fate of Chemicals in the Aquatic Environment*; Dickson, K.L., Maki, A.W., Cairns, J., Jr., Eds.; Ann Arbor Science, Ann Arbor, MI, **1982**, 325-346.

18. Larson, R.J.; Rothgeb, T.M.; Shimp, R.J.; Ward, T.E.; Ventullo, R.M. *J. Am. Oil Chem. Soc.* **1993**, *70*, 645.

19. Swisher, R.D. *Surfactant Biodegradation, 2nd Ed.*; Marcel Dekker, New York, NY, **1987**, 1085.

20. Schoberl, P. *Tenside Surfactants Detergents* **1989**, *26*, 86.

21. Larson, R.J.; Bishop, W.E. *Soap/Cosmetics/Chemical Specialties* **1988**, *April*, 58.

22. Divo, C.; Cardini, G. *Tenside Detergents* **1980**, *17*, 30.

23. Giger, W.; Alder, A.C.; Brunner, P.H.; Marcomini, A.; Siegrist, H. *Tenside Surfactants Detergents* **1989**, *26*, 95.

24. Thurman, E.M.; Barber, L.B., Jr.; LeBlanc, D.R. *J. Contam. Hydrol.* **1986**, *1*, 143.

25. Thurman, E.M.; Willoughby, T.; Barber, L.B., II; Thorn, K.A. *Anal. Chem.* **1987**, *59*, 1798.

26. Field, J.A.; Leenheer, J.A.; Thorn, K.A.; Barber, L.B., II; Rostad, C.; Macalady, D.L.; Daniel, S.R. *J. Contam. Hydrol.* **1992**, *9*, 55.

27. Field, J.A. *Unpublished Ph.D. Thesis* ; Colorado School of Mines, Golden, CO, **1990**, 207.

28. Bolsman, T.A.B.M.; Veltmaat, F.T.G.; van Os, N.M. *J. Am. Oil Chem. Soc.* **1988**, *65*, 280.

29. Hand, V.C.; Williams, G.K. *Environ. Sci. Technol.* **1987**, *21*, 370.

30. Hand, V.C.; Rapaport, R.A.; Pittinger, C.A. *Chemosphere* **1990**, *21*, 741.

31. Di Toro, D.M.; Dodge, L.J.; Hand, V.C. *Environ. Sci. Technol.* **1990**, *24*, 1013.

32. Schwarzenbach, R.P.; Westall, J.C. *Environ. Sci. Technol.* **1981**, *15*, 1360.

33. Chiou, C.T.; Porter, P.E.; Schmedding, D.W. *Environ. Sci. Technol.* **1983**, *17*, 227.

34. LeBlanc, D.R. *U.S. Geological Survey Water-Supply Paper 2218* **1984**, 27 p.

35. Barber, L.B., II; Thurman, E.M.; Schroeder, M.P.; LeBlanc, D.R. *Environ. Sci. Technol.* **1988**, *22*, 205.

36. Ceazan, M.L.; Thurman, E.M.; Smith, R.L. *Environ. Sci. Technol.* **1989**, *23*, 1402.

37. Harvey, R.W.; Smith, R.L.; George, L. *Appl. Environ. Microbiol.* **1984**, *48*, 1197.

38. Harvey, R.W.; Barber, L.B., II. *J. Contam. Hydrol.* **1992**, *9*, 91.

39. Metge, D.W.; Brooks, M.H.; Smith, R.L.; Harvey, R.W. *Appl. Environ. Microbiol.* **1993**, *59*, 2304.
40. Smith, R.L.; Duff, J.H. *Appl. Environ. Microbiol.* **1988**, *54*, 1071.
41. LeBlanc, D.R.; Garabedian, S.P.; Hess, K.M.; Gelhar, L.W.; Quadri, R.D.; Stollenwerk, K.G.; Wood, W.W. *Water Res. Res.* **1991**, *27*, 895.
42. Garabedian, S.P.; LeBlanc, D.R.; Gelhar, L.W.; Celia, M.A. *Water Res. Res.* **1991**, *27*, 911.
43. Harvey, R.W.; George, L.H.; Smith, R.L.; LeBlanc, D.L. *Environ. Sci. Technol.* **1989**, *23*, 51.
44. Harvey, R.W.; Kinner, N.E.; Smith, R.L.; LeBlanc, D.R. *Water Res. Res.* **1993**, *29*, 2713.
45. Smith, R.L.; Howes, B.L.; Duff, J.H. *Geochim. Cosmochim. Acta* **1991**, *55*, 1815.
46. Smith, R.L.; Howes, B.L.; Garabedian, S.P. *Appl. Environ. Microbiol.* **1991**, *57*, 1997.
47. Kent, D.B.; Davis, J.A.; Anderson, L.C.D.; Rea, B.A.; Waite, T.D. *Water Res. Res.* **1994**, *30*, 1099.
48. Smith, R.L.; Harvey, R.W.; LeBlanc, D.R. *J. Contam. Hydrol.* **1991**, *7*, 285.
49. Field, J.A.; Miller, D.J.; Field, T.M.; Hawthorne, S.B.; Giger, W. *Anal. Chem.* **1992**, *64*, 3161.
50. Freyberg, D.L. *Water Res. Res.* **1986**, *22*, 2031.
51. Roberts, P.V.; Goltz, M.N.; Mackay, D.M. *Water Res. Res.* **1986**, *22*, 2047.
52. Barber, L.B., II *Environ. Sci. Technol.* **1994**, *28*, 890.
53. Smith, R.L.; Duff, J.H. *Appl. Environ. Microbiol.* **1988**, 54, 1071.
54. Jannasch, H.W.; Pritchard, P.H. *Mem. Inst. Ital. Idorobiol. Suppl.* **1972**, *29*, 289.
55. Bengtsson, G. *Microb. Ecol.* **1989**, *18*, 235.
56. Harvey, R.W.; George, L. *Appl. Environ. Microbiol.* **1987**, *53*, 2992.
57. Kinner, N.E.; Bunn, A.L.; Harvey, R.W.; Warren, A.; Meeker, L.D. *U.S. Geol. Survey, Water Resources Invest. Rept. 91-4034,* **1991**, 141.

RECEIVED January 11, 1995

Chapter 9

Effect of Different Surfactant Concentrations on Naphthalene Biodegradation

James R. Mihelcic[1], Dan L. McNally[1], and Donald R. Lueking[2]

[1]Department of Civil and Environmental Engineering and
[2]Department of Biological Sciences, Michigan Technological University,
Houghton, MI 49931

Addition of surfactants has been proposed to assist remediation of environmental systems contaminated with hydrophobic organic chemicals. Growth studies and degradation tests are presented to help improve interpretation of how surfactants may assist or inhibit biodegradation. Growth studies confirmed that the test organism does not utilize the nonionic surfactants, Triton X-100 or Brij 58, as carbon sources. Growth studies also demonstrated what surfactant concentrations would inhibit growth of the test organism on glycerol. A concentration of Triton X-100 up to 12% and Brij 58 up to 5% inhibited growth of the test organism by approximately 25%. It was observed that the rate of naphthalene degradation was not enhanced or inhibited by a Triton X-100 concentration of 0.0012% (v/v). This concentration was determined to be below the CMC. However, a Triton X-100 concentration of 1.2% enhanced biodegradation.

Hydrophobic organic chemicals (HOCs), such as polycyclic aromatic hydrocarbons, are widely distributed in the environment. After being discharged to an environment such as the subsurface, HOCs will partition to several phases such as the sorbed phase and separate phase nonaqueous phase liquids. One concern about the fate of HOCs in subsurface environments is that they may not be readily available for natural or engineered biodegradation after they have partitioned into either of these two phases (*1*).

To overcome this problem of bioavailability, several researchers have examined the addition of surfactants to assist the biodegradation of HOCs. These studies have typically been conducted with synthetic surfactants (*2-4*), though some researchers have used purified surfactants of biological origin (*5,6*). The studies with synthetic surfactants usually use nonionic surfactants because gram negative bacteria are known to be much more resistant to them compared to ionic surfactants (*7*). This is because of the net negative charge displayed by both gram negative

0097–6156/95/0594–0112$12.00/0

and gram positive organisms. Typically, these studies have reported observations of experiments conducted with different HOCs and surfactants at various concentrations. The observations are sometimes made without examining mechanisms which result in the stimulatory or inhibitory effect of the surfactants on biodegradation. Consequently, some of these studies have not provided insight into the specific mechanism(s) which cause surfactants to assist or inhibit biological processes. For example, while greater than CMC concentrations of Triton X-100 (0.05 to 1.0%) inhibited phenanthrene mineralization and a concentration of 0.01%, near the CMC, slightly inhibited phenanthrene mineralization (2), greater than CMC concentrations of the surfactant, Neodol, stimulated growth on decane and tetradecane (3).

Some explanations have either been identified or hypothesized to explain the inhibiting effects of surfactants on biological processes. These include toxicity of the surfactant to the test organism (8,9), use of the surfactant as a preferred growth substrate (9), prevention of cell attachment to an organic liquid phase (10,11) and sequestering of solute into micelles (2). And in several cases, the addition of a purified biosurfactant has been shown to inhibit biodegradation (12,13) or exhibit antimicrobial effects (14,15). The literature also suggests that in some instances cell attachment is a prerequisite for biodegradation of an organic separate phase. For example, a decrease in the lag phase and subsequent higher total growth, as measured by a protein assay, has been reported for growth on hexadecane in the presence of 0.01% Triton X-100 and Brij 35 (16). It was believed that this level of surfactant permitted cell/hexadecane contact and thus facilitated biodegradation. However, in a different study, 0.1% Triton X-100 was shown to prevent attachment of cells to heptamethylnonane which contained the substrate, hexadecane (11). Finally, studies with rhamnolipids, a biosurfactant, showed that they could increase the hydrophobicity of cells which initially had low hydrophobicity. This resulted in enhanced octadecane degradation (6).

Although the knowledge base of how surfactants enhance or inhibit biodegradation of HOCs has expanded greatly in the past few years, additional studies are still required which provide information on the specific mechanisms (either biological or physical) which cause a particular observation. Accordingly, the purpose of this work was to demonstrate a methodology to study the effect of a synthetic surfactant on biodegradation of an HOC like naphthalene. It is hoped that this methodology will provide insight into some of the biological mechanisms which cause surfactants to inhibit, enhance, or have no effect on the biodegradation of an organic substrate.

Experimental Section

Chemicals. All chemicals were reagent grade or better. Naphthalene and glycerol were obtained from Fisher Scientific Co. (Pittsburgh, PA), Triton X-100 (Octyl phenoxy polyethoxyethanol) and Brij 58 (polyoxyethylene 20 cetyl ether) were obtained from Sigma Chemical Company (St. Louis, MO). Sigma reports that Triton X-100 has a CMC of 0.014% and average molecular weight of 628 while Brij 58 has a CMC of 0.00078% and average molecular weight of 1,120. Basal salts medium (BSM) was used in all biological studies and consists of the following salts (per liter of distilled water): 4 g K_2HPO_4, 4 g Na_2HPO_4, 2 g $(NH_4)_2SO_4$, 0.2 g $MgSO_4 \cdot 7 \cdot H_2O$, 0.001 g $FeSO_4 \cdot 7H_2O$, and 0.001 g $CaCl_2 \cdot 2H_2O$. The pH of BSM medium is 7. All medium and glassware were sterilized by autoclave at 120°C, 15 psi, for 20 min prior to use.

Test Organism. Designated strain Uper-1 was obtained via enrichment culture from soil samples acquired from a coal storage site located in Michigan's Upper Peninsula. It was subsequently identified as a gram negative rod *Pseudomonas fluorescens* by the use of an API NFT strip developed by Biomerieux (l'Etoile, France). Uper-1 is routinely prepared and maintained in BSM with 1.5% (w/v) glycerol. Uper-1 also possesses the ability to oxidize naphthalene. Stock cultures are maintained at -20°C in a BSM and glycerol medium adjusted to 30% (w/v) glycerol.

Evaluation of Critical Micelle Concentration. The critical micelle concentration (CMC) of systems containing Uper-1 in BSM and the surfactants, Triton X-100 and Brij 58, was determined by the ring method to measure surface tension, using a Du Nouy interfacial tensiometer No. 70545 (CSC Scientific Company, Inc. Fairfax, VA). This method consists of placing different concentrations of surfactant and approximately 3×10^8 cells per ml of BSM into a sterilized, disposable 60x15-mm plastic petri dish (Fisher Scientific Co.). A 6-cm platinum-iridium ring is immersed approximately 5 mm into the liquid sample and the ring is then raised. When the film breaks, a reading is recorded in dynes/cm. This is the apparent force exerted on the ring which is corrected to the true surface tension according to the tensiometer manufacturer's guidelines. The effect of BSM and cells on surface tension was also determined for BSM concentrations ranging from 0 to 100% and Uper-1 concentrations of 0 to 10^{10} cells/ml. All tests were conducted at 20°C.

Evaluation of Biological Effects of Triton X-100 and Brij 58. Previous researchers have suggested that synthetic surfactants, such as Triton X-100 and Brij 58, may inhibit biodegradation of an HOC by either serving as a competitive growth substrate or by being toxic. Accordingly, tests were conducted to demonstrate if Triton X-100 and Brij 58 could serve as growth substrates and to determine the surfactant concentration that would not cause inhibition of the test organism.

To determine if Triton X-100 and Brij 58 could be used as a growth substrate by Uper-1, the streak plate method was utilized. The surfactants would serve as the sole source of carbon to sustain growth. Uper-1 was initially grown on 1.5% glycerol in BSM and brought to log phase. After washing the residual glycerol, Uper-1 was streaked out onto petri plates containing BSM, 1.5% agar, and 0.1% (v/v) surfactant. Plates were incubated at 30°C for 72 hr. The absence of growth was determined by visual inspection of the plates. Uper-1 growth on Triton X-100 and Brij 58 was also determined by inoculating a broth medium containing 10 ml BSM and 0.1% (v/v) surfactant in 18-ml culture tubes. These samples were monitored for changes in turbidity using a Klett Summerson Colorimeter for 24 hr. Controls, containing no glycerol, and blanks, containing no cells, were also run to identify any effects other than those caused by the surfactants.

Tests were also conducted to determine what concentrations of Triton X-100 and Brij 58 may inhibit or stimulate the growth of Uper-1. The growth of Uper-1 in the presence of various concentrations of surfactant below and above the CMC was examined. The procedure consisted of adding an extensively washed inoculum of Uper-1, grown to log phase, in a BSM and 1.5% glycerol broth in 125-ml side arm flasks. Samples consisted of 10 ml BSM, 1.5% glycerol, 500 ul of cells, and surfactant. The addition of the 500 ul of cells resulted in an initial Klett reading of 30 which corresponds to approximately 3×10^8 cells/ml. Triton X-100 was studied at concentrations ranging from 0.0007 to 12% (v/v) while Brij 58 was studied at concentrations ranging from 0.00002 to 5% (v/v). After preparation, side-arm flasks were placed in a shaking water bath set at 30°C for 6 hours. Uper-1 growth was monitored by recording the Klett readings hourly. A control was prepared without surfactants for comparison purposes. A blank with a known surfactant concentration was used to calibrate the Klett Summerson Colorimeter for the sample with the corresponding concentration. The collected data were regressed to determine the specific growth rate for Uper-1 when growing on glycerol in the presence of different surfactant concentrations.

Degradation of Naphthalene in the Presence of Triton X-100. The effect of various concentrations of Triton X-100 on the biodegradation of naphthalene was examined. Uper-1 was initially grown on 1.5% glycerol in BSM, brought to log phase growth, and washed 4 times to ensure the removal of any residual glycerol. A solution containing BSM and Triton X-100 was added to 15-ml centrifuge tubes so that the Triton X-100 concentrations to be tested were below (0.0012%) and above (1.2%) the CMC. Triton X-100 was selected because it is commonly used in literature, easy to handle, and has a relatively high CMC which corresponds to a higher monomer concentration which is more convenient when looking at the influence of monomers and micelles. Then 100 ul of a N,N-dimethyl formamide (DMF)-naphthalene stock solution was added to give an initial naphthalene concentration of approximately 3 mg/L in each tube. DMF has been shown in our laboratories, at amounts used in this study, to not inhibit the growth of Uper-1. Each tube was inoculated by adding 100 ul BSM containing enough Uper-1 to achieve a final cell concentration of 4.5×10^8 cells/ml. All samples were prepared in triplicate. Tubes were immediately sealed with screw caps which contained Teflon-lined septa. Samples were placed in a tumbler rotating at 20 rpm in the dark at 20°C. Controls included the DMF-naphthalene solution and the two concentrations of Triton X-100. The controls were used to examine the possible adverse effects of the surfactant on the efficiency of the extraction method which was employed prior to naphthalene analysis. Blanks included the DMF-naphthalene solution only and were used to monitor the abiotic loss of naphthalene. A 0.1% (v/v) concentration of sodium azide was added to the controls and blanks to prevent any microbial growth.

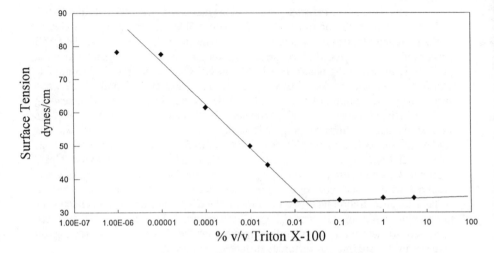

Figure 1a. Surface tension versus Triton X-100 concentration. The critical micelle concentration (CMC) was determined to be 0.012% (v/v).

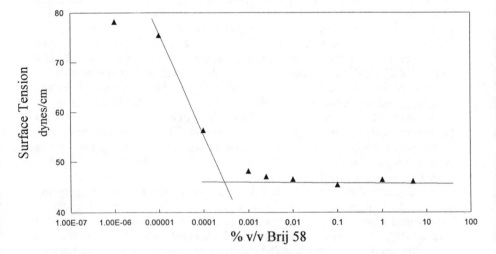

Figure 1b. Surface tension versus Brij 58 concentration. The critical micelle concentration (CMC) was determined to be 0.00045% (v/v).

Samples were periodically removed from the tumbler and prepared for analysis. The extraction procedure followed a modified Bligh and Dyer extraction method as described by Ames (*17*). This method uses chloroform and methanol as the extracting solvents. The ratio of 1/2/0.8 (v/v/v) of chloroform/methanol/sample was used. The extraction was conducted using 40-ml sample vials with screw caps and Teflon-lined septa. After final mixing, both phases were allowed to clarify for 10 min before an aliquot of the PAH-chloroform mixture were placed in 1.8-ml GC vials for analysis. The analysis for the degradation of naphthalene was conducted using a GC-MS with modified EPA Method 625. A mass spectra was obtained using a Hewlett Packard model 5890 gas chromatograph interfaced with a Hewlett Packard model 5870 mass spectrometer via a single-stage jet separator. The column was a DB-5 capillary column 30-m long, 0.25-mm ID, and a 0.25-um film thickness (J+W Scientific, Folsom, CA). The injection and detector ports' temperatures were 275°C. The carrier gas was helium and was set at a flow of 1 ml/min. The column temperature initially was 65°C and was ramped 25°C per min to a final temperature of 140°C, and ramped again at 10°C per min to a final temperature of 270°C. The detector (ion source) temperature was 250°C. The MS was in single ion monitoring mode with 100 scans per minute and an ionization current of 200 eV greater than the autotune setting. Minimum detection level for naphthalene was 1 ug/L.

Results and Discussion

The determination of the CMC of the Uper-1, BSM, and surfactant system allowed the examination of the influence on the CMC of adding microbial cells and BSM. It also assisted selection of the surfactant concentrations, below and above the CMC, used in later studies. Figure 1a shows the surface tension for different Triton X-100 concentrations which resulted in a CMC of 0.012%. Figure 1b shows the surface tension for different Brij 58 concentrations which resulted in CMC of 0.00045%. A 0.012% Triton X-100 concentration corresponds to 0.2 mM assuming an average molecular weight of 628 while a 0.00045% Brij 58 concentration corresponds to 0.004 mM assuming an average molecular weight of 1120. It was observed that the presence of a high salt concentration and high concentration of cells had no significant effect on the measured CMC of our experimental systems. That is, our values are similar to those reported by the supplier. Additionally, it was proven that neither various cell nor BSM concentrations appreciably influenced the surface tension (data not shown).

Some nonionic surfactants are reported to biodegrade (*18*). Consequently, they may be a preferred substrate and thus, limit or inhibit the biodegradation of HOCs. Accordingly, tests to determine if Uper-1 could utilize either Triton X-100 or Brij 58 as growth substrates showed that neither surfactant could support growth. This was verified by noting the absence of growth on plates after 72 hr and no increase in turbidity of surfactant solutions after 24 hr. Therefore, the use of surfactants as a preferential growth substrate would not have a negative influence on the degradation of naphthalene in later studies which combined a surfactant and naphthalene.

Inhibition effects were evaluated for various concentrations of Triton X-100 and Brij 58. Klett unit readings were recorded to monitor growth. The data were analyzed by plotting on semilog graph paper Klett units versus time. The doubling time (g), or time to double the turbidity from about 40 to 80 Klett units, was obtained and then the doubling time was related to the specific growth rate, u, by the relationship, $u = 0.693/g$.

Table I shows the specific growth rate as a function of surfactant concentration. These results show cell growth was observed at all surfactant concentrations tested. However, there generally appears to be slight inhibition for Triton X-100 concentrations above 0.0015% after which the inhibition remains constant at approximately 25% below the growth rate obtained in the absence of Triton X-100. For Brij 58 slight inhibition was observed at all surfactant concentrations examined and resulted in a specific growth rate approximately 25% below the growth rate obtained in the absence of Brij 58. These results suggest that surfactant concentrations as high as 12% for Triton X-100 and 5% for Brij 58 can be used in biodegradation studies without significantly decreasing the growth rate of Uper-1. It was observed for Uper-1 that the toxic effect of both surfactants did not increase with higher surfactant concentrations. This suggests that the effect is limited to the surfactant monomer in this situation. However, this may not be true with all surfactants and may be a function of items such as the CMC, organism, and surfactant structure. Finally, the two nonionic surfactants did not exert an appreciable toxic effect at the concentrations employed in this study on our test organism in studies examining naphthalene biodegradation. Several researchers have shown that above CMC levels of surfactants are required to significantly enhance the aqueous solubility of hydrophobic organic chemicals (19,20). In our study, it is clear that when using pure microbial cultures, easy-to-perform growth studies can be utilized to determine apparent surfactant toxicity to a test organism and whether a surfactant can be used as a competitive carbon source.

Table I. Specific Growth Rate of Uper-1 as a Function of Triton X-100 and Brij 58 Concentrations

Triton X-100 (% v/v)	u (hr⁻¹)	Brij 58 conc. (% v/v)	u (hr⁻¹)
0	0.31	0	0.24
0.00073	0.33	0.00002	0.16
0.0015	0.25	0.00005	0.12
0.0061	0.20	0.00007	0.16
0.014	0.22	0.01	0.21
0.060	0.28	0.05	0.19
0.60	0.23	0.10	0.20
6.0	0.24	0.50	0.18
12	0.20	1.0	0.18
		5.0	0.20

Other studies performed in our laboratories with a different microorganism are much more dramatic with regards to surfactant inhibition of growth. A PAH-degrading microorganism was isolated from soil obtained from a hazardous waste site in California. It has been identified as *Pseudomonas stutzeri* and designated as strain W-2. Table II shows the results of growth studies with W-2 with the same BSM-1.5% glycerol medium used in studies with Uper-1. These results show a specific growth rate of 0.27 hr^{-1} in the absence of Triton X-100. However, at 1% and 5% Triton X-100, significant inhibition of the test organism occurs. In fact, the specific growth rate in the absence of surfactant is reduced by approximately 44% and 67% for 1% and 5% Triton X-100 concentrations, respectively. Concentration of surfactants in this range has been used in many biodegradation studies because as mentioned previously it is accepted that these concentrations are required to enhance solubility of organic solutes, especially in the presence of soil. In addition, Triton X-100 concentrations of 1% have been found to be toxic to the yeast, *Candida lipolytica*, growing on dodecane while 0.01% enhanced oxygen consumption during growth on dodecane (*21*). Thus it is seen that the various microbial strains respond differently to the same surfactant.

**Table II. Specific Growth Rate of W-2
as a Function of Triton X-100 Concentration**

Triton X-100 (% v/v)	u (hr^{-1})
0.0	0.27
0.1	0.29
0.5	0.25
1.0	0.15
5.0	0.09

Note that in Tables I and II, the lowest Triton X-100 concentration tested resulted in a very slight stimulation of the growth for both Uper-1 and W-2. While not a significant increase in the specific growth rate for this situation, we have observed stimulation of microbial growth at low surfactant concentrations. Another study showed in column experiments that the mineralization of phenanthrene and biphenyl was enhanced by 10 mg/L of the nonionic Novel II surfactant (CMC of approximately 50 mg/L) (*4*). In that study, the authors provide no explanation for this observation. They noted that the lower surfactant concentration enhanced mineralization without promoting solute desorption from the soil. One potential explanation for the enhanced biodegradation at lower surfactant concentrations may be an increase in the permeability of the cell membrane. This is especially important because the classical method for transport of hydrophobic chemicals through cellular membranes is believed to be by diffusion, which is considered a passive transport mechanism (*22*). While the inner leaflet of the outer membrane in gram negative organisms is composed of a phospholipid bilayer, the outer leaflet is

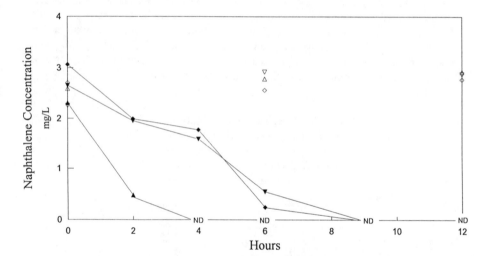

Figure 2. Naphthalene degradation by *Pseudomonas fluorescens* strain designated Uper-1 in the presence of 0.0012% (v/v) and 1.2% (v/v) Triton X-100.
(▲1.25% (v/v) surfactant with cells, ▼ 0.00125% (v/v) surfactant with cells,
♦ no surfactant with cells, △ 1.25% (v/v) surfactant without cells, ∇ 0.00125%
(v/v) surfactant without cells, ◊ no surfactant without cells, ND non-detect;
detection limit for naphthalene was 0.001 mg/L)

composed of lipopolysaccharide (LPS). The LPS has polysaccharide side chains which extend out into the external medium. This makes the exterior of gram negative organisms hydrophilic which presents a formidable barrier for penetration of the outer membrane by hydrophobic chemicals. However, disruption of the outer membrane by addition of surfactants can reduce the efficiency of this barrier in preventing or minimizing transport of HOCs across the outer membrane.

Experiments were conducted with Triton X-100 to investigate the influence of nonionic surfactant concentrations below and above the CMC on the biodegradation of naphthalene. The previous biological studies had determined that if inadequate biodegradation occurred when using Uper-1, it most likely would not be due to toxic concentrations of Triton X-100 or preferential use of Triton X-100 as a growth substrate. Thus, any observed inhibition of naphthalene degradation might be due to a phenomenon such as physical entrapment of substrate inside surfactant micelles.

Figure 2 shows the naphthalene concentration over time in the presence of Triton X-100 concentrations, below and above the CMC. Controls which contained no Uper-1 remained constant over the time course of the experiment. This indicates the surfactant did not lower the efficiency of the extraction and lose naphthalene at the interface or in the water-methanol phase. Blanks which contained no surfactant or Uper-1 also remained constant which demonstrates no significant loss of naphthalene occurred from abiotic loss mechanisms. The results clearly show that a Triton X-100 concentration of 0.0012% had no significant effect on the rate of naphthalene biodegradation while a Triton X-100 concentration of 1.2% enhanced the removal of naphthalene. In the presence of the above CMC levels of the surfactant, aqueous naphthalene concentration decreased from 3 mg/L to less than detection levels in less than 4 hr while with below the CMC level of surfactant and the absence of surfactant, naphthalene concentration decreased from 3 mg/L to nondetectable levels in greater than 6 hr. These results suggest that in our experimental system, the one surfactant concentration tested below the CMC did not influence naphthalene biodegradation. That is, no enhancement or inhibition of Uper-1 occurred at sub CMC concentrations of surfactant.

However, the one surfactant concentration tested which was above the CMC of Triton X-100 enhanced naphthalene biodegradation. It has been observed that phenanthrene mineralization was inhibited in the presence of above CMC levels of Triton X-100 (2). That study hypothesized, but never showed conclusively, that this result may have been from micellular solubilization of the phenanthrene which made it unavailable for biodegradation. A parallel study, using fluorescence monitoring, demonstrated reduction in biodegradation of naphthalene in the presence of 0.2% Triton X-100 (23). That study hypothesized that the naphthalene was either sequestered in micelles or the surfactant inhibited biodegradation.

It is difficult to compare bioavailability of micelle associated HOCs with different surfactant/solute mixtures. Solubilization of an organic solute by micelles is a complex process. For example, solubilization of HOCs is believed to be a function of many items, including the lipid/hydrophilic balance of the surfactant (*i.e.*, ethylene oxide chain length) (24). In addition, it is believed that different organic solutes localize themselves in different parts of the micelle. For example, ethylbenzene may situate itself in the hydrocarbon interior of micelles while anthracene and naphthalene may only penetrate deeply into the palisade layer (25). In addition, solutes may also only adsorb to the micellular surfaces or experience short penetration into a palisade layer. The location of the solute within the micelle may be important because surfactants may facilitate biodegradation by either allowing contact and subsequent solute diffusion from micelles to the outer membrane or perhaps by a fusion between the surfactant and the cell membrane which may be required for uptake of organic solute. Thus, the location of solute in micelles may explain the discrepancy of some studies using similar surfactants but different solutes.

This study demonstrated easy-to-perform methods to eliminate the reasons of surfactant toxicity and surfactant use as a preferential growth substrate observed for inhibition of HOC biodegradation in the presence of surfactants. This study also showed that a 1.2% concentration of the nonionic surfactant, Triton X-100, enhanced naphthalene biodegradation. Current investigations in our laboratory are attempting to isolate the mechanism(s) which result in enhanced biodegradation in the presence of surfactants. As engineers attempt to develop more efficient methods for remediation of contaminated subsurface and aquatic systems, they will require additional information on how surfactants of chemical or biological origin enhance or inhibit biodegradation.

Acknowledgments

This work was supported by the National Science Foundation under Grant No. BCS-9110136 and a U.S. Department of Education Fellowship awarded to D.L. McNally. The Government has certain rights to this material. Margaret Sottile and Rebecca Tarnowski assisted with the growth studies. The soil sample which W-2 was isolated from was provided by Gary McGinnis of Michigan Technological University's Institute of Wood Research.

Literature Cited

1) Mihelcic, J.R.; Lueking, D.R.; Mitzel, R.J.; Stapleton, J.M. Biodegrad. 1993, 4, 141-153.

2) Laha, S.; Luthy, R.G. Environ. Sci. Technol. 1991, 25, 1920-1930.

3) Bury, S.T.; Miller, C.A. Environ. Sci. Technol. 1993, 27, 104-110.

4) Aronstein, B.N.; Alexander, M. Appl. Microbiol. Biotechnol. 1993, 39, 386-390.

5) Obermeyer, A.; Muller-Hurtig, R.; Wagner, F. Appl. Microbiol. Biotechnol. 1990, 32, 485-489.

6) Zhang, Y.; Miller, R. Appl. Environ. Microbiol. 1994, 60, 2101-2106.

7) Vaara, M. Microbiol. Rev. 1992, 56, 395-411.

8) Cserhati, T.; Illes, Z.; Nemes, I. Appl. Microbiol. Biotechnol. 1991, 35, 115-118.

9) Tiehm, A. Appl Environ. Microbiol. 1994, 60, 258-263.

10) Mimora, A.; Watanabe, S.; Takeda, I. *J. Ferment. Technol.* 1971, 49, 255-271.

11) Efroysom, R.A.; Alexander, M. *Appl. Environ. Microbiol.* 1991, 57, 1441-1447.

12) Hommel, R.; Stuwer, O.; Weber, L., Kleber H.P. *14th Int. Congr. Biochem. Abstracts.* 1988, 11, 242.

13) Foght, J.M.; Gutnick, D.L., Westlake, D.W.S. *Appl. Environ. Microbiol.* 1989, 55, 36-42.

14) Haferburg, D.; Hommel R.; Claus R.; Kleber, H.P. *Adv. Biochem. Engrg. Biotechnol.* 1986, 33, 53-93.

15) Lang, S.; Katsiwela, E.; Wagner, F. *Fat Sci. Technol.* 1989, 91, 363-366.

16) Breuil, C.; Kushner, D.J. *Can. J. Microbiol.* 1979, 26, 223-231.

17) Ames, G.F. *J. Bacteriol.* 1968, 95, 833-843.

18) Schick, M.J. In *Nonionic Surfactants*; Schick, M.J., Marcel Dekker, Inc., Yew York, New York, 1967, pp. 971-996.

19) Kile, D.E.; Chiou, C.T. *Environ. Sci. Technol.* 1989, 23, 832-838.

20) Edwards, D.A.; Luthy, R.G.; Liu Z. *Environ. Sci. Technol.* 1991, 25, 127-133.

21) Whitworth, D.A.; Moo-Young, M.; Viswanatha, T. *Biotechnol. Bioengrg.* 1973, 15, 649-675.

22) Nikaido, H. *Biochem. Biophys. Acta* 1976, 433, 118-132.

23) Putcha, R.V.; Domach, M.M. *Environ. Progress*, 1993, 12, 81-85.

24) Nakagawa, T. In *Nonionic Surfactants*; Schick, M.J., Marcel Dekker, Inc., Yew York, New York, 1967, pp. 558-603.

25) Riegleman, S.; Allowala, N.A., Hrenoff, M.K.; Strait, L.A. *J. Col. Sci.* 1958, 13, 208-217.

RECEIVED December 13, 1994

Chapter 10

Influence of Anionic Surfactants on Bioremediation of Hydrocarbons

Joseph D. Rouse[1], David A. Sabatini[1,3,4], and Jeffrey H. Harwell[2,3]

[1]School of Civil Engineering and Environmental Science,
[2]School of Chemical Engineering and Materials Science, and
[3]Institute for Applied Surfactant Research, University of Oklahoma,
Norman, OK 73019

Manometric respirometers were used to evaluate the influences of a wide range of surfactants (mostly anionic) on microbial activity in systems containing a hydrocarbon substrate (naphthalene), basic nutrients, and an activated sludge seed. Sulfated surfactants served as preferred substrates in the presence of naphthalene. Sulfonated surfactants, however, did not readily serve as substrates and demonstrated various degrees of enhancement and inhibition of naphthalene oxidation. Testing with a series of twin-head group anionic surfactants (diphenyl oxide disulfonate--DPDS) with varying straight chain hydrocarbon tail lengths (C6 through C22) indicated that microbial oxidation rates of naphthalene are generally enhanced at surfactant levels above the critical micelle concentration (CMC) with the surfactants that have mid range tail lengths (C10 to C16). With sodium dodecylbenzene sulfonate (SDBS), oxidation of naphthalene was largely suppressed at supra CMC levels. Using a dextrose substrate, slight increases in lag times and decreases in oxidation rates were observed with DPDS surfactants above and below their CMCs. With SDBS, however, lag times were decreased and dextrose oxidation rates were enhanced at supra-CMC surfactant concentrations.

The use of surfactants in attempts to enhance environmental remediation efforts has been of considerable interest in recent years (1-6). Surfactants, or surface active agents, tend to migrate to surfaces and lower interfacial tensions. This often results in the formation of emulsions, thus greatly increasing the surface area between immiscible liquids. Another characteristic of surfactants is their tendency to form micelles or aggregates of surfactant molecules when the critical micelle concentration (CMC) is surpassed. The micellar core constitutes a hydrocarbon

[4]Corresponding author

pseudo-phase capable of solubilizing hydrocarbon compounds into solution to levels above their water solubilities (1,7-9).

These characteristics have lead to the consideration of using surfactants when biodegradation of poorly soluble compounds is of concern (10-15). However, with the addition of surfactants, both enhancements and inhibitions of biodegradation for organic compounds have been reported and the mechanisms involved when surfactant-substrate-microorganism interactions occur are not adequately understood. In an effort to better understand the factors involved in such systems, a lengthy review of the literature addressing this subject has been completed (16); a summary of findings from this review is presented below.

In systems containing mixed microbial cultures, biodegradation of hydrocarbons was inhibited by the use of nonionic surfactants at levels above the CMC (13,14,17). Also, cationic and anionic commercial surfactants have been noted for their damaging effects on cell membranes (11,18-20) and for this reason they are often disregarded for use in biological systems. Some anionic surfactants, though, are considered to be biodegradable (15,21,22,23); thus, it is expected that these surfactants would not be harmful to microbial cultures under certain conditions. Various anionic surfactants with this potential have recently demonstrated efficient solubilization of hydrocarbons with low losses due to precipitation and sorption (23,24). It is hypothesized that these high performance anionic surfactants may, under proper environmental conditions, prove to be compatible with a treatment scenario that includes biodegradation.

While surfactants at concentrations above the CMC can greatly enhance the apparent water solubility of hydrocarbon compounds, the actual aqueous component of the compound--apart from the micellar pseudo-phase--may be greatly reduced under nonequilibrium conditions or when the hydrocarbon excess phase is depleted (25,26). If a microorganism's route of hydrocarbon substrate uptake is strictly via the aqueous phase, this concentration reduction along with hydrocarbon exit rates from micelles could potentially be controlling factors. However, the potential for interactions between commercial surfactants and cell membranes has been a point of discussion in recent literature (12,13,18,20) and it has been suggested that micellar contents are directly accessible to membrane bound enzymes in the degradation pathway (27). In the latter scenario, steric or ionic compatibility of surfactants with membrane components, rather than substrate accessibility, would be of greater concern. These factors and others must be considered when evaluating surfactant influences on biodegradation of organic compounds.

The objective of this research is to investigate the influence of various anionic surfactants on the microbial oxidation of a hydrocarbon substrate. This objective is met by comparing oxidation rates of naphthalene in microbial systems containing various types and concentrations of surfactants to control samples. To better understand the factors involved in the biodegradation assays, partitioning of hydrocarbon from surfactant solutions into an alkane phase will be evaluated; partition rates will serve as a potential indicator of bioavailability of solubilized hydrocarbon. Also, surfactant toxicity assays will be conducted using a soluble substrate that is not responsive to micellar solubilization. Finally, sorption of surfactants into biomass will be quantified so as to evaluate the propensity for surfactants to become incorporated into microbial membranes and thus directly interact with the organisms.

Materials and Methods

The straight chain alkyl diphenyl oxide disulfonate (DPDS) surfactants used in this research were from the DOWFAX series as supplied by Dow Chemical Co. (Midland, MI) and consisted of [R] C6L (DPDS-C6), [R] 3B2 (DPDS-C10), XU 40568.00 (DPDS-C12), [R] 8390 (DPDS-C16), and XUR 1528-9301953-4 (C20-24, say DPDS-C22), where hydrocarbon chain lengths are designated by "-C#". The molecular weights, CMCs, and chemical formulas of these surfactants are shown in Table I. With the exception of DPDS-C22, about 20% of the surfactants by weight are double tailed and about 10% mono sulfonated--DPDS-C22 is strictly mono tailed and 98% disulfonated. The surfactants were received in liquid form at about one molar concentration with NaCl levels at 0.2% max with the exception of DPDS-C22 which was at about 0.5 molar concentration and 3.0% max NaCl.

Table I. Anionic surfactants used in this research that did not serve readily as substrates in respirometeric analyses.

surfactant	MW	CMC	design mol formula
DPDS-C6	474	7 mM [a]	$C_6H_{13}C_{12}H_7O(SO_3Na)_2$
DPDS-C10	542	6 mM [a]	$C_{10}H_{21}C_{12}H_7O(SO_3Na)_2$
DPDS-C12	575	5 mM [a]	$C_{12}H_{25}C_{12}H_7O(SO_3Na)_2$
DPDS-C16	642	3 mM [b]	$C_{16}H_{33}C_{12}H_7O(SO_3Na)_2$
DPDS-C22	682	1 mM [a]	$C_{22}H_{45}C_{12}H_7O(SO_3Na)_2$
SDBS (C12)	348	4 mM [a]	$C_{12}H_{25}C_6H_4SO_3Na$

[a] Estimated by capillary rise method (24).   [b] Estimated from surface tension data developed by a maximum bubble pressure method (21).

Sodium dodecyl benzene sulfonate (SDBS) was purchased from Aldrich Chemical Co. (Milwaukee, WI) and sodium lauryl (dodecyl) sulfate (SDS), from Fisher Scientific Co. (St. Louis, MO), both in dry form. Samples of sodium lauryl ether sulfates, CS-130 and CS-330 (Steol series with 1 and 3 ethoxylate units per surfactant molecule, respectively), were received from the Stepan Co. (Northfield, IL) and mono- and di-sulfated sorbitol ester surfactants (C12 alkyl sorbital ester with 20 ethoxylate units) were received from Lonza Inc. (Long Beach, CA). T-MAZ 20 was purchased from PPG Ind. Inc. (Gurnee, IL) and Triton X-100 (tert-octyl, branched alkyl phenol with ca. 9.5 ethoxylate units) from Amersham (Arlington Heights, IL). All the above surfactants were used without further purification. Naphthalene (99% purity, Aldrich Chemical Co.), a common environmental pollutant, was used as the hydrocarbon substrate of interest.

A Beckman System Gold chromatograph (Beckman Instruments, Inc., San Ramon, CA) was used for high-performance liquid chromatography (HPLC) analyses to quantify naphthalene and DPDS surfactants at a UV wavelength of 276 nm and SDBS at a UV of 225 nm. A mobile phase of 85% methanol was used at a flow rate of 0.7 mL/min with a 250 X 4.0 mm Spherisorb C18 reverse-phase column (Scientific Glass Engineering Pty. Ltd., Ringwood, Australia).

Respirometer assays were conducted at room temperature (ca. 22°C) with Hach (Loveland, CO) manometric apparatuses (model 2173B) using Hach BOD nutrient buffer (pillows) and lithium hydroxide (powder pillows) to remove carbon dioxide. The stock naphthalene solutions that served as the hydrocarbon substrate source were prepared by mixing naphthalene crystals (in excess) in deionized water in 4 L amber bottles with magnetic stirrers. HPLC analyses of six stock solutions determined the soluble naphthalene concentrations to be 34.4 mg/L (standard deviation of 3.8 mg/L). The minimum mixing time evaluated was 5 days. For use as a substrate, 200 mL aliquots of this stock were added by graduated cylinder to 500 mL amber assay bottles containing appropriate volumes of surfactant solution, dilution water, and nutrients. The microbial seed was added last to yield a final volume of 420 mL. With the above dilution, the final naphthalene concentration was about 16.4 mg/L which yields a theoretical BOD of 49.1 mg/L. Exposure to the atmosphere during loading was kept to a minimum to prevent volatilization losses of naphthalene. For surfactant toxicity assays, the stock solution consisted of a mixture of dextrose and glutamic acid (84 mg/L each) which was diluted (100 mL stock to 420 mL total) to a theoretical BOD of 40 mg/L for use as a substrate.

During assays, samples were stirred gently with magnetic stir bars at about 200 rpm. For each concentration of surfactant tested, duplicate samples were run with and without naphthalene (or the dextrose-glutamic acid substrate) and for each series tested, four surfactant-free samples and duplicate endogenous controls (only nutrients and microbial seed in deionized water) were executed.

The microbial seed stocks were obtained from a nearby activated sludge domestic wastewater treatment facility. Sludge samples from the aeration basin were decanted to a suspended solids concentration (re. Standard Methods (28), Sect. 2540 D) of 5.0 g/L. A 2 mL aliquot of this seed was added to each assay sample (ca. 0.5% v/v). The prepared seed stocks were stored at 4°C and were generally used over a two week period before being replaced. A sludge seed derived in this manner should be reasonably reproducible, thus reducing testing variables.

Cumulative oxygen uptake or biochemical oxygen demand (BOD) versus time was used as the standard means of quantifying microbial activity. Determinations of enhancements and inhibitions of the influence of surfactants on the biodegradation of naphthalene as compared to the response in surfactant-free systems were based on three items: (i) time prior to onset of oxygen uptake (lag time), (ii) approximate zero order rate of oxygen uptake during about the first 30 hours of activity after the lag time (r squared regression coefficients generally greater than 0.95), and (iii) first order BOD "k" constant determined over the entire period of oxygen uptake beyond the lag time (a larger "k" value denotes a more rapid reaction or greater affinity between the components of the reaction). The ultimate oxygen demand for BOD "k" determinations was determined by Langmuirian regression (weighted for high concentrations) as opposed to the theoretical stoichiometric demand (which was not achieved). With the Langmuirian input to the BOD "k" calculations, r squared values were usually 0.98 or greater.

Partitioning assays of naphthalene between surfactant solutions and an alkane phase were conducted with 10 mL of aqueous surfactant solution and 10 mL of decane in 22 mL capped glass vials. Components were added carefully to avoid formation of emulsions and stirred gently at about 200 rpm. At all surfactant concentrations, initial naphthalene concentrations were designed to be near saturation. Sampling for HPLC analysis of naphthalene was done with a micro-syringe and only required about 40 uL of decane per sample event (including ca. 20 uL to rinse the syringe prior to sampling). The extracted decane was diluted and mixed into 2-propanol (usually 2.0 mL) and then injected into the HPLC. Naphthalene standards were diluted into similar volumes of 2-propanol.

Sorption of DPDS surfactants onto biomass was evaluated by mixing surfactant solutions with activated sludge on a wrist action shaker for 8 hours, centrifuging for 20 minutes, and measuring the equilibrium surfactant in the supernatant. Surfactant concentrations were determined by a methylene blue spectrophotometric assay (Standard Methods (28), Sect. 5540 C); UV or conductivity methods could not be used due to excessive interference from the biomass as evidenced in surfactant free controls. Reagent and sample volumes were reduced proportionately from the procedural directive to reduce waste generation. Preliminary assays verified that sorption was at equilibrium after 8 hours with samples at 4, 8, and 12, hours yielding the same results. The activated sludge had been settled, decanted and rinsed with deionized water twice before being adjusted to a suspended solids concentration of 12 g/L. This sludge was then diluted with surfactant solution in 34 mL glass vials to a final sludge concentration of 10 g/L. Respirometric assays were conducted with these surfactants and biomass at the same concentrations (without nutrients) which demonstrated that no significant oxygen uptake was occurring within 12 hours, thus assuring that significant biodegradation of surfactants was not occurring.

Results and Discussion

Preliminary Analyses. Respirometric assays were run to establish testing conditions that would render the microbial utilization of the hydrocarbon substrate to be the limiting or controlling factor for oxygen uptake rate analysis. Various concentrations of naphthalene (ranging from about 8 to 24 mg/L) were tested and oxygen utilization rates were shown to increase with increasing substrate concentration. This indicates a non-zero order region of oxygen utilization with respect to substrate concentration. Thus, rate determinations using the finalized 16.4 mg/L naphthalene concentration should be sensitive to environmental factors that affect the microbial population directly or restrict their access to the substrate. Also, tests performed with and without stirring and with pure oxygen headspace indicated no variation in oxygen utilization rates thus indicating that oxygen transfer was not limiting the biological reaction. Similar assays done with a dextrose substrate demonstrated that stirred samples had higher oxygen consumption rates than non-stirred samples, thus indicating that for these systems with the more labile substrate, oxygen transfer may be a controlling factor (especially under stagnant conditions).

With each set of experiments, surfactant-free controls were run. The mean values of key parameters determined from these controls with their standard deviations in brackets are as follows: lag time, 41.6 hrs (4.9); initial zero order rate, 0.374 mg/L*hr (0.052); ultimate BOD (Langmuirian plateau), 43.3 mg/L (8.6); first order BOD "k" (base 10), 0.00346 hr-1 (0.00067) (n-1, where n=50; generally ca. 20 data points per each "n"). Subsequent experimental analyses will be based on a normalization with respect to the surfactant-free controls performed at the time of each experiment and not the overall mean values above. These values, however, can be multiplied by the reported normalized data to make reasonable estimates of their actual values. The normalized values of the above means would, of course, all be 1.000 and the normalized standard deviations would be the given deviation divided by its corresponding mean.

The rubber cups (for holding lithium hydroxide) and stirring bars used in the respirometer assays retained a strong aromatic odor after exposure to naphthalene. Therefore, after each use, they were cleaned by soaking in methanol (ca. 6 hr) and then water (ca. 6 hr) prior to air drying. The extent to which a volatile hydrocarbon partitions between water and headspace air and materials of the apparatus could be a contributing factor to the performance of a system. The precise impact of this on oxidation rates would be difficult to determine; thus, the importance of pursuing more precise experimental methods (e.g., quantifying substrate disappearance and evolution of degradation by-products by chromatographic techniques, use of radio labeled compounds, etc.) to advance our understanding of environmental systems are evident.   However, a respirometric method, as used here, can be useful for screening large numbers of samples to determine where further detailed study would be most beneficial.

Tests with Sulfated (and Nonionic) Surfactants. Results of respirometric assays conducted with sulfated anionic surfactants (data not shown) indicated clearly that all of these surfactants served readily as substrates (see Materials and Methods for listing of these surfactants).   While yielding useful information concerning the biodegradability of the surfactants themselves, interpretation of their influence on the degradation of the intended hydrocarbon substrate--the intent of this study--is difficult. Differences in oxygen uptake responses between samples with and without naphthalene may not be valid indications of the fate of the separate components due to interactions between the diverse microbial populations in the presence of the multiple substrates involved. Clearly, further work would be necessary to quantify the substrate directly rather than indirectly as with respiration. For comparative purposes, similar test where also run using the common nonionic surfactants T-MAZ 20 and Triton X-100; these surfactants also revealed their susceptibility to serve readily as substrates for the microbial seed used here.

Systems containing Lonza and the nonionic surfactants with naphthalene addition displayed higher cumulative oxygen uptakes than the surfactant-only or naphthalene-only assays. This suggests that the hydrocarbon substrate is being utilized. In the systems containing Steol surfactants or SDS, the oxygen demands

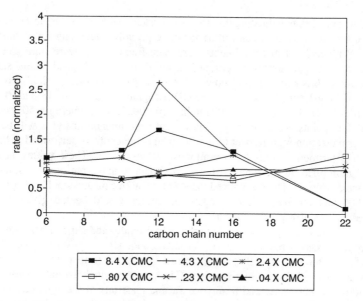

Figure 1. Summary of initial approximate zero order oxygen uptake rates for the DPDS series with a naphthalene substrate. DPDS surfactants are indicated by carbon chain number. Surfactant concentrations are normalized to CMCs (see text).

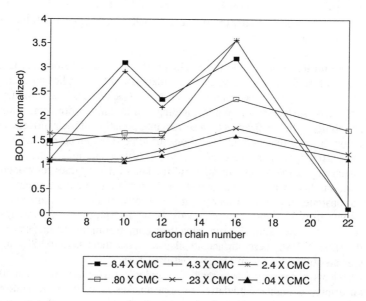

Figure 2. Summary of first order BOD "k" oxygen uptake rates for the DPDS series with a naphthalene substrate. DPDS surfactants are indicated by carbon chain number. Surfactant concentrations are normalized to CMCs (see text).

were generally the same with or without naphthalene over the period of testing which would suggest that these surfactants were serving as a much preferred substrate over naphthalene and perhaps exerting some toxicity toward naphthalene degrading organisms.

The labile nature of the sulfated and nonionic surfactants studied here would be a point of concern when their use in environmental remediation efforts is considered even if enhancement in degradation of a hydrocarbon contaminant could be displayed in laboratory studies.   In an environment like the subsurface, oxygen could be rapidly depleted while degrading the surfactant, thus adding reaeration and surfactant replenishment expenses to a remediation project. The potential for bio-plugging, as well, would be increased. However, with the use of alternate electron acceptors, the response of organisms in contaminated environments containing these surfactants could be quite different; this could be an avenue for further research (29).

Test with Sulfonated Surfactants. For the sulfonated surfactants, appreciable exertions of oxygen demand in naphthalene-free samples were not evidenced during the normal 5 to 7 day testing periods with BOD responses generally less than that of the endogenous controls which typically had five day BOD values of 5 to 10 mg/L. This overall lack of degradation of these surfactants under testing conditions made them good candidates for use in respirometric assays to evaluate their influence on biodegradation of organic compounds.

Assays conducted with SDBS at concentrations approximating and surpassing the CMC frequently displayed no significant oxidation of naphthalene during their periods of testing (see Table II). To check for possible interferences that could be responsible for fowling these systems, nutrients at the concentration used in the respirometric assays and SDBS at concentrations ranging from approximately 0.1 to 40 mM were mixed in clear glass flasks at room temperature and observed after a few hours of gentle stirring. The nutrient buffer mixture went completely into solution in all samples except at 1 mM SDBS, which remained cloudy. This may have been due to some form of phase separation which might account for the failure of assays conducted near this surfactant concentration to respond. At higher SDBS concentrations, however, the frequent failure to respond could be related to surfactant interferences with cell metabolic processes or the restriction of the hydrocarbon substrate in a poorly accessible solubilized state (discussed in following sections). The occasional positive naphthalene oxidation response (e.g., at ca. 9 and 30 mM) is probably not a surfactant concentration-specific phenomenon. Instead, it may be evidence of specific microbial species adapting to a harsh environment, in which case, a well defined microbial seed might have generated more consistent responses.

A summary of the influence of the DPDS surfactant series on the microbial oxidation of naphthalene is depicted in Figures 1, 2, and 3 (derived from data in reference 30). In these figures, surfactant concentrations are grouped and normalized with respect to CMC values (average concentration of each group shown in legends) and oxygen uptake rates and lag times are normalized to results of surfactant-free controls. As shown in Figures 1 and 2, the zero and first order rates tend toward enhancement in samples with surfactant levels above the CMC for DPDS-C10, -C12, and -C16. The normalized rate values for the DPDS-C6 surfactant systems demonstrated little variation from unity, while for DPDS-C22 slight enhancements in oxygen uptake rates below the CMC and inhibition above the CMC were observed. The responses associated with supra-CMC values for the C22 surfactant are less conclusive due to a lack of data in the 2.4 X CMC bracket; also, where greatly inhibited responses occurred, there is the possibility that a positive response may have occurred with a longer period of testing. Below the CMC, for the whole range of DPDS surfactants there is little variation from unity for rate values, except for the assays containing DPDS-C16. This case shows a trend toward enhancement with increasing surfactant concentration.

Table II. Effects of SDBS on microbial oxidation of naphthalene. Testing periods ranged from 139 to 261 hours. Each entry represents averaged results from two samples.

SDBS conc.	approx. zero order rate	BOD "k"	maximum BOD (plateau)	lag time
mM	normalized to	surfactant--	free sample	values
0.10	1.038	1.013	1.044	0.722
0.23	1.022	1.267	0.647	0.774
0.50	0.719	1.106	0.628	1.366
0.91	(*)	(*)	(*)	(*)
2.00	(*)	(*)	(*)	(*)
4.57	(*)	(*)	(*)	(*)
9.14	0.851	1.526	0.474	1.550
10.0	(*)	(*)	(*)	(*)
20.0	(*)	(*)	(*)	(*)
30.0	1.700	2.264	0.528	2.098
40.0	(*)	(*)	(*)	(*)

(*) = no significant response within period of testing.

Increases in lag time prior to onset of degradation for the DPDS series (Figure 3) do not appear to be correlated with CMC values. Assays with DPDS-C10 and -C12 show no significant variation in trend over the range of surfactant concentrations tested, with the C12 systems being close to unity (i.e., no significant inhibition of lag time). Conversely, the results of assays with the C6 and C16 surfactants are difficult to explain; both systems demonstrated an increase in lag time with increasing surfactant concentrations followed by a decrease, which does not relate to CMC values. Others have made similar observations for these relatively short and long tailed DPDS surfactants (i.e., C6 and C16)--difficult to explain trends evidenced for various abiotic parameters with respect to surfactant concentration (31). Within the DPDS series, only the experiments involving DPDS-C22 (not included in Figure 3) demonstrated a consistent failure to respond within the period of testing when surfactant levels were clearly above the CMC.

Partitioning of Naphthalene from DPDS Solutions into Decane. In an attempt to better evaluate the factors involved in the biodegradation assays, the kinetics of partitioning for naphthalene from aqueous surfactant solutions into a separate alkane phase (decane) were investigated. Partition rates of naphthalene into the decane phase could serve as an indicator of the bioavailability of the solubilized hydrocarbon. With the less than saturated concentration of naphthalene used in these respirometric assays (ca. 16.4 mg/L), a 30 mM addition of DPDS-C16 would reduce the strictly aqueous component of the hydrocarbon (i.e., that which is not solubilized into surfactant micelles) to about 1.2 mg/L (see reference 26 for calculation procedure). The importance of investigating the accessibility of solubilized hydrocarbon to microorganisms is thus evident.

Experiments were conducted with DPDS-C6, -C10, and -C16 and SDBS with surfactant concentrations ranging up to 50 mM. In all cases surfactant was never detected in the decane phase; this is attributed to the high water solubilities of these anionic surfactants. These results differ with those by others that showed a significant transfer of the nonionic Triton X-100 from aqueous solution into hexane (14). This agrees with the relative sorption potentials of these types of surfactants (24).

The partitioning of naphthalene out of the SDBS and DPDS-C6 solutions into decane were nearly the same as that observed from naphthalene saturated water without surfactant--about 78% of the naphthalene appeared in the decane within the first hour and 90% within the first 2 hours. For the DPDS-C16 systems, however, slower rates were evidenced, and increasing surfactant concentrations were correlated with decreasing rates of naphthalene transfer as shown in Figure 4. DPDS-C10 also influenced rates of transfer with the results being intermediate (by surfactant concentration) to those of the -C16 and -C6 systems (qualitatively assessed). The first order rate of naphthalene transfer (in the form of a BOD "k", base 10) for DPDS-C16 at 30 mM is 0.189 hr-1 which is almost two orders of magnitude higher than the first order rate for microbial oxidation of naphthalene in surfactant-free systems (k = 0.00346 hr-1). These results indicate that solubilization capacity, as indicated by surfactant hydrocarbon tail length and micellar surfactant concentration, does appear to affect the exit rates of hydrocarbons from micellar surfactant solution into an alkane phase, and thus potentially into a microbial membrane as well. The effect, however, is not enough to limit biodegradation activity. Similar conclusions were made by Laha and Luthy (14) using systems that involved the transfer of phenanthrene from a nonionic surfactant solution (Triton X-100) into hexane.

These results do not address the longest tailed surfactant in the series, DPDS-C22, which is associated with inhibition of oxygen utilization rates at supra-CMC levels; additional product samples of this experimental surfactant were not available for further testing. It is not likely, however, that micellar exit rates of solubilized

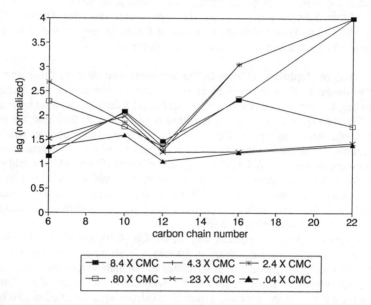

Figure 3. Summary of lag time prior to onset of oxygen uptake for the DPDS series with a naphthalene substrate. Surfactant concentrations are normalized to CMCs (see text).

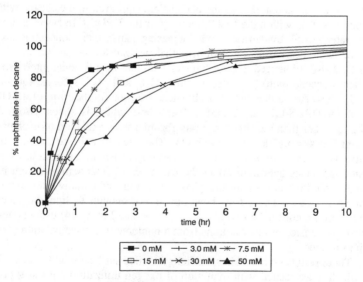

Figure 4. Transfer of naphthalene from DPDS-C16 surfactant solutions into decane.

hydrocarbon or any other partitioning phenomena associated with this surfactant would be of such a magnitude as to inhibit biodegradation. It has been suggested, though, that some lipid micelles inhibit the release of solubilized hydrocarbons and thus, potentially, inhibit biodegradation (32).

Direct Interactions of Surfactants with Microorganisms. Thus, enhancements in microbial oxidation rates of naphthalene (an already water soluble substrate) by the addition of certain DPDS surfactants at supra-CMC levels can not be explained by micellar exit rates, hydrocarbon phase transfer, or solubilization potential. If the level of naphthalene being used were toxic, then a reduction in the strictly aqueous component due to micellar solubilization might serve to lower the effective concentration to a more amiable level. However, preliminary studies demonstrated that the naphthalene concentration being used was not inhibitory, i.e., higher concentrations resulted in higher oxidation rates. Also, the DPDS surfactants themselves were evidently not serving as substrates. This leads to the consideration that the surfactants may be interacting with cell envelopes in a way that results in enhanced rates of substrate transport to enzymes. Increased liquidity of membranes by low levels of surfactants has been suggested as a potential reason for observed enhancements in microbial activity (17). Perhaps something of this nature could occur at higher surfactant concentrations as well.

If direct incorporation of surfactants into cell membranes is occurring, the steric or conformational compatibility of a surfactant with membrane components would be a potential concern. Perhaps the relatively small sized DPDS-C6 is a poor match for the larger cell membrane lipids. This could explain the poorer results associated with the C-6 surfactant versus the longer tailed DPDS surfactants which could conceivably do a better job of blending into the membrane. By this logic, however, the DPDS-C22 systems should also have demonstrated enhancements in microbial oxidation rates. It should be noted, though, that the experimental DPDS-C22 is not of the same sulfonation and alkylation ratios as the other surfactants in the series and may contain other additives which make comparisons of results with other systems difficult.

If the observed enhancements and inhibitions are a result of surfactants interacting with cell wall or membrane components (as opposed to altering the substrate state and thus its bioavailability), assays using a soluble substrate that is not responsive to micellar solubilization (i.e., is not solubilized into the micellar pseudo-phase) could be informative. Accordingly, biodegradation assays were run using a dextrose-glutamic acid substrate with DPDS surfactants and SDBS at concentrations below and above their CMCs. The results of these respirometric assays using DPDS-C6, -C10, and -C16 are shown in Figures 5, 6, and 7, respectively. Interestingly, these results indicate that the DPDS surfactants do influence the biodegradation of the dextrose-glutamic acid substrate. Lag times were increased by surfactant addition in all cases by about 2 to 10 hours--generally similar to results with the naphthalene substrate. Furthermore, at all surfactant concentrations tested, DPDS-C6 (Figure 5) and DPDS-C10 (Figure 6) systems had oxygen uptake rates (approximate zero order) that were significantly lower (by a 95% confidence interval) than those of surfactant-free controls. For DPDS-C16 (Figure 7), though,

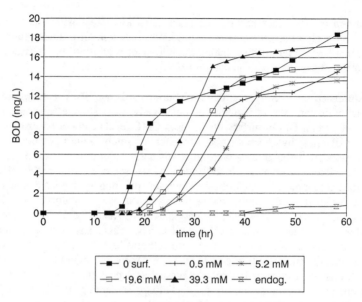

Figure 5. Cumulative BOD for microbial assays with DPDS-C6 and a
dextrose-glutamic acid substrate. Each datum point is an averaged value from
duplicate assays.

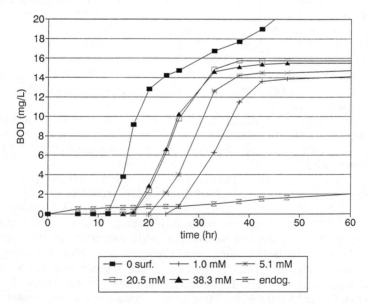

Figure 6. Cumulative BOD for microbial assays with DPDS-C10 and a
dextrose-glutamic acid substrate. Each datum point is an averaged value from
duplicate assays.

oxidation rates at all surfactant concentrations were not significantly different from those of controls.   These results give credence to the idea that shorter tailed surfactants are less amiable to microbial membranes.

For the assays conducted with SDBS and the dextrose-glutamic acid substrate, a very low surfactant concentration (0.2 mM) had no influence on oxidation rate as compared to surfactant-free controls (Figure 8).   For supra-CMC levels (20 and 39 mM), however, lag times were decreased about 5 hours and initial zero order rates were significantly enhanced.   With 1.0 mM SDBS the oxidation response was greatly inhibited.   This corroborates the previously noted system fowling at this SDBS concentration.   The frequent failure for SDBS systems to respond at supra-CMC levels with the naphthalene substrate (see Table II), even though the exit rates of this hydrocarbon out of SDBS solutions appeared rapid (see previous section), gives credence to the possibility that some form of interaction is occurring between the surfactant and membrane bound enzymes in the degradation pathway.

To evaluate the propensity of the DPDS surfactants to interact with or become incorporated into microorganisms, experiments were attempted to quantify the sorption of these surfactants into biomass.   DPDS-C6, -C10, and -C16 at 7.5 mM were mixed, individually, with activated sludge (10 g/L) and the subsequent equilibrium concentrations of the surfactants were quantified.   Results counter-intuitively indicated that the most hydrophobic C16 surfactant is least susceptible to sorb into the biomass and, conversely, the least hydrophobic C6 is the most prone to sorb.   Perhaps the more hydrophobic surfactants put more energy into micelle formation and thus are less prone to interfere with microorganisms.   These results could have been influenced by surfactant-induced liberation of compounds from the biomass that interfere with surfactant analysis and cause artificially high measurements.   Surfactant-free controls, however, did not display any such interference.   As an approximation, the C6, C10, and C16 surfactants appeared to sorb out of solution at about 0.15, 0.07, and 0.04 g of surfactant per g of biomass, respectively, which would suggest that some significant interaction is occurring. These results are considered to be preliminary; ongoing work is continuing to asses this phenomenon.

Summary and Conclusions

Respirometric assays demonstrated that sulfated surfactants served readily as substrates with a mixed microbial culture.   Sulfonated surfactants, however, did not appreciably serve as substrates and their addition to cultures with a naphthalene substrate demonstrated varying degrees of enhancement and inhibition of oxygen uptake.   A series of twin-head disulfonated surfactants with varying hydrocarbon tail lengths indicated that microbial oxidation rates of naphthalene are generally enhanced at surfactant levels above the CMC when mid-length (C10 to C16) surfactants are used.   With the anionic surfactant SDBS, microbial oxidation of naphthalene was largely suppressed at supra-CMC levels.   Partitioning rates of naphthalene from surfactant solutions into decane were rapid compared to oxygen uptake rates, suggesting that solubilized hydrocarbon is readily available to microorganisms. Results of assays with a dextrose substrate and sorption experiments of surfactants

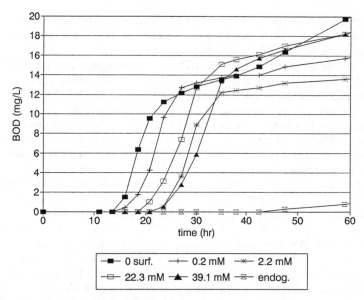

Figure 7. Cumulative BOD for microbial assays with DPDS-C16 and a dextrose-glutamic acid substrate. Each datum point is an averaged value from duplicate assays.

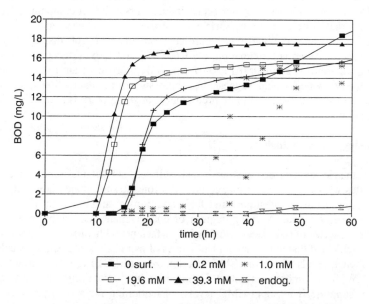

Figure 8. Cumulative BOD for microbial assays with SDBS and a dextrose-glutamic acid substrate. Each datum point is an averaged value from duplicate assays except for 1.0 mM SDBS systems which are singular.

onto biomass indicated that direct interaction of surfactants with microorganisms may be a significant factor influencing the biodegradation of organic compounds. These results thus indicate that surfactant system design (i.e., surfactant structure, concentration, etc.) is critical to successful enhancement of bioremediation.

Literature Cited

(1)   Harwell, J.H. *In Transport and Remediation of Subsurface Contaminants*; Sabatini, D.A., Knox,R.C., Eds.; ACS Symposium Series 491; American Chemical Society: Washington, DC, 1992; pp 124-132.

(2)   West, C.C. *In Transport and Remediation of Subsurface Contaminants*; Sabatini, D.A., Knox,R.C., Eds.; ACS Symposium Series 491; American Chemical Society: Washington, DC, 1992; pp 149-158.

(3)   Edwards, D.A.; Laha, S.; Liu, Z.; Luthy, R.G. *In Transport and Remediation of Subsurface Contaminants*; Sabatini, D.A., Knox,R.C., Eds.; ACS Symposium Series 491; American Chemical Society: Washington, DC, 1992; pp 159-168.

(4)   Palmer, C.; Sabatini, D.A.; Harwell, J.H. *In Transport and Remediation of Subsurface Contaminants*; Sabatini, D.A., Knox,R.C., Eds.; ACS Symposium Series 491; American Chemical Society: Washington, DC, 1992; pp 169-181.

(5)   Fountain, J.C. *In Transport and Remediation of Subsurface Contaminants*; Sabatini, D.A., Knox,R.C., Eds.; ACS Symposium Series 491; American Chemical Society: Washington, DC, 1992; pp 182-191.

(6)   Palmer, C.D.; Fish, W. *Chemical Enhancements to Pump-and-Treat Remediation*; EPA/540/S-92/001; Environmental Protection Agency: Washington, D.C., 1992.

(7)   Shiau, B.J.; Sabatini, D.A.; Harwell, J.H. *Ground Water.* 1994, 32, 561-569.

(8)   West, C.C.; Harwell, J.H. *Environ. Sci. Technol.* 1992, 26, 2324-2330.

(9)   Rosen, M.J. *Surfactants and Interfacial Phenomena*; John Wiley & Sons, Inc.: New York, 1989.

(10)   Bury, S.J.; Miller, C.A. *Environ. Sci. Technol.* 1993, 27, 104-110.

(11)   Lupton, F.S.; Marshall, K.C. *Geomicrobiol. J.* 1979, 1, 235-247.

(12)   Guerin, W.F.; Jones, G.E. Appl. *Environ. Microbiol.* 1988, 54, 937-944.

(13)   Laha, S.; Luthy, R.G. *Environ. Sci. Technol.* 1991, 25, 1920-1930.

(14)   Laha, S.; Luthy, R.G. *Biotechnol. Bioeng.* 1992, 40, 1367-1380.

(15)   Tiehm, A. *Appl. Environ. Microbiol.* 1994, 60, 258-263.

(16)   Rouse, J.D.; Sabatini, D.A.; Suflita, J.M.; Harwell, J.H. *CRC Crit. Rev. Environ. Sci. Technol.* 1994, accepted for publication.

(17)   Van Hoof, P.L.; Rogers, J.E. in *Biosystems Technology Development Program. Bioremediation of Hazardous Waste*; EPA/600/R-92/126; U.S. EPA: Washington, DC, 1992; pp 105-106.

(18)   Ulitzur, S.; Shilo, M. *J. Gen. Microbiol.* 1970, 62, 363-370.

(19)   Shabtai, Y.; Gutnick, D.L. *Appl. Environ. Microbiol.* 1985, 49, 192-197.

(20)   Swisher, R.D. *Surfactant Biodegradation,* In Surfactant Science Series 18;
       Marcel Dekker, New York, 1987.
(21)   Dow Chemical Co., Midland, MI 48674.
(22)   Stepan Co., Northffield, IL 60093.
(23)   Rouse, J.D.; Sabatini, D.A.; Brown, R.E.; Harwell, J.H. Evaluation of
       Ethoxylated Alkylsulfate Surfactants for use in Subsurface Remediation,
       submitted to *Water Environ. Research.* 1994.
(24)   Rouse, J.D.; Sabatini, D.A.; Harwell, J.H. *Environ. Sci. Technol.* 1993, 27,
       2072-2078.
(25)   Christian, S.D.; Smith, G.A.; Tucker, E.E.; Scamehorn, J.F. *Langmuir.*
       1985, 1, 564-567.
(26)   Rouse, J.D.; Sabatini, D.A.; Deeds, N.E.; Brown, P.E.; Harwell, J.H.
       Effects of Unsaturated Hydrocarbon Concentrations on Anionic Surfactant-
       Enhanced Remediation as Evaluated by Semi-Equilibrium Dialysis, submitted
       to *Environ. Sci. Technol.* 1994.
(27)   Miller, R.M.; Bartha, R. *Appl. Environ. Microbiol.* 1989, 55, 269-274.
(28)   *Standard Methods for the Examination of Water and Wastewater,* 18th
       Edition; APHA, AWWA, WEF; Washington DC, 1992.
(29)   Hutchins, S.R.; West, C.C.; Wilson, B.E. Preliminary Studies on Surfactant
       Enhancement of Nitrate-Based Bioremediation. In Book of Abstracts, 207th
       ACS National Meeting, San Diego, CA. March 13-17, 1994.
(30)   Rouse, J.D. Ph.D. Dissertation, University of Oklahoma, Norman, OK,
       1994.
(31)   Loughney, T., Personal correspondence: Dow Chemical Co., Midland,
       MI 48674.
(32)   Falatko, D.M.; Novak, J.T. *Water Environ. Res.* 1992, 64, 163-169.

RECEIVED December 1, 1994

IMPLEMENTATION ISSUES AND PROCESS OPTIMIZATION

Chapter 11

Lessons from Enhanced Oil Recovery Research for Surfactant-Enhanced Aquifer Remediation

G. A. Pope[1] and W. H. Wade[2]

[1]Department of Petroleum and Geosystems Engineering and
[2]Department of Chemistry and Biochemistry, University of Texas,
Austin, TX 78712

This paper provides a brief summary of some key elements of surfactant enhanced oil recovery technology that are relevant to the application of the use of surfactants to the remediation of contaminated soils. Surfactant screening and laboratory testing, the use of polymers and foams for mobility control, the use of tracers for characterization and performance assessment and modeling and field testing are discussed and important lessons from EOR research and experience are identified for each of these topics. Surfactant enhanced aquifer remediation (SEAR) technology could benefit tremendously from these EOR lessons, which were learned the hard way over more than thirty years.

Although some research using alkaline agents to generate surfactants from crude oil and other very early research in the oil industry goes back many decades, most published surfactant enhanced oil recovery research dates from 1963, the year petroleum sulfonates were patented for this use. Research activity was particularly high during the decade of the 1970s, when thousands of papers were published on the use of surfactants to improve oil recovery. Hundreds of field tests of this process have been conducted during the past 30 years. Many of these field tests, particularly those done after 1980, were technical successes, although we learn as much from the failures sometimes as the successes. Eventually, this research on hundreds of surfactants and mixtures of surfactants led to the identification of surfactants that could be effectively used under very harsh conditions. Thus, there is a huge body of scientific and practical data on the use of surfactants in the subsurface. No attempt will be made to summarize this enormous body of data here, which would require at least a book length document, but rather only a few items that seem particularly pertinent to surfactant enhanced aquifer remediation will be briefly discussed. The principles involved in both applications are the same, yet there are significant differences in the conditions, criteria for success, costs and so forth that must be kept in mind in this discussion. Introductions to surfactant EOR can be found in Lake (*1*), Pope and Bavière (*2*) and Bourrel and Schechter (*3*). No attempt will be made to reference the SEAR literature here since it can be found elsewhere in this proceedings. Suffice it to say that there are several high quality research efforts underway to develop and test this promising remediation technology.

0097–6156/95/0594–0142$12.00/0
© 1995 American Chemical Society

There are many other elements of enhanced oil recovery besides the surfactant behavior that are pertinent to SEAR and these also will be briefly mentioned. For example, the characterization of the oil reservoir by the use of tracers before the surfactant is injected as well as other methods of characterizing the formation have been found to be very useful. There have been far more failures of surfactant EOR due to poor characterization or operation of the reservoir than due to problems with the surfactant. From almost the beginning, it was found that mobility control was necessary for the efficient use of surfactants to recovery crude oil from reservoirs. By far the most common method of doing this is the use of high molecular weight water-soluble polymers. These polymers increase the viscosity of both the injected surfactant solution and the water used to displace the surfactant solution (called a slug since it is typically only a small fraction of the total pore volume of the reservoir). The most important benefit of this thickening of the water is the attenuation of the permeability variations of the formation that cause bypassing or inefficient sweep of the low permeability zones and this benefit is highly pertinent to SEAR. Other benefits include the elimination of hydrodynamic instabilities (fingers), decreased retention of the surfactant in the formation and greater control of where the injected fluid goes into the formation. Each of these benefits will be discussed below. Another aspect of enhanced oil recovery that applies to SEAR as well as to other remediation processes is performance assessment technology. One of the most important elements of any process in the subsurface is the evaluation of how well the process worked and this is a very difficult task requiring special technology that is conceptually and operationally similar for EOR and SEAR.

Surfactant EOR is based primarily on the reduction of interfacial tension (IFT) to mobilize residual oil saturation. When water is injected into oil reservoirs, the capillary forces are very large compared to the viscous forces, and as a result about 20 to 40% oil saturation is trapped as ganglia held by the strong capillary forces. Typically, the IFT between the oil and water must be reduced by three or four orders of magnitude to mobilize all of this residual oil saturation, i.e., to on the order of 0.001 mN/m. The precise requirements depend on the rock characteristics, which can be represented by a function of the capillary number defined as

$$N_c = \frac{\left| k \cdot \nabla \Phi_w \right|}{\sigma},$$

where k is the permeability, Φ is the potential, and σ is the interfacial tension. A capillary desaturation curve measured by Delshad (4) is shown in Fig. 1. Capillary numbers on the order of 0.01 were required to completely displace the residual oil saturation from this Berea sandstone. As shown by Morrow and Songkran (5), buoyancy forces can also effect the mobilization of residual oil and can be expressed by the Bond number, which is defined by

$$N_B = \frac{g k \left(\rho_w - \rho_o \right)}{\sigma},$$

where ρ_w and ρ_o are the densities of the water and oil, respectively. The buoyancy effect is usually considered insignificant in EOR applications, but there are exceptions. On the other hand, it is very important under typical aquifer conditions in SEAR applications. The Bond number is typically higher for surfactant remediation of aquifers than for surfactant flooding of oil reservoirs because the permeability of the aquifer is typically higher and the density difference between the aqueous and organic phases will be higher for DNAPLs such as PCE and TCE. Mobilization of the NAPL will occur for high Bond numbers even if the capillary number is small. When the risk of vertical migration of DNAPL is unacceptable, then this is a disadvantage since it requires increasing the IFT to reduce the Bond number. On the other hand, if mobilization is acceptable, then it is an advantage since it makes mobilization easier and more complete. This is because the viscous and gravity forces reflected in the capillary and Bond numbers both promote mobilization of the trapped NAPL. Recent column experiments at both the University of Texas and the University of Michigan have verified the importance of the Bond number when surfactant displaces DNAPL in either the vertical or horizontal orientation of the column.

Only a very small amount of surfactant is required to reduce the IFT and displace the oil from a core when the IFT is ultralow and the capillary number is high. The principle limitation is the retention of surfactant by adsorption, trapping or other mechanisms. Thus, as little as 10% of a pore volume of a 3% surfactant solution may be adequate to displace 90% of the oil from the core. This same surfactant might recover only 60% of the oil from a reservoir because of bypassing of much of the oil due to reservoir heterogeneities, well pattern effects and other factors unrelated to the effectiveness of the surfactant on a microscopic scale. The criteria for SEAR is very different. First, we would like to recover as close as possible to 100% of the organic contaminant rather than say 60% as in EOR. Secondly, we are likely to inject multiple pore volumes of surfactant (with possible recycling after one pore volume) rather than only a small fraction of a pore volume. Thus, surfactant retention in the formation is likely to be much less important for SEAR than for EOR. The final amounts of the contaminant are likely to be in the water and on the surface of the soil as its concentration approaches zero. These play no role in EOR at all.

An even more significant difference between EOR and SEAR is that we may purposefully not want to mobilize at all with SEAR, but rather only solubilize. This will depend on many factors some of which are still poorly known. We are much more likely to choose mobilization as the primary recovery mechanism if the contaminant is a light nonaqueous phase liquid (LNAPL) than if it is a dense nonaqueous phase liquid (DNAPL). Even if it is a DNAPL, we may safely mobilize it if we can show that it is confined and will not vertically migrate once the IFT is lowered. The solubilization process is much less efficient than mobilization via low IFT, so there is a great incentive to define the conditions under which each is appropriate. The amount of organic that a surfactant solution solubilizes is related to the reduction of the IFT by the Chun-Huh (6) equation

$$\sigma = \frac{c}{S^2},$$

where σ is the interfacial tension, S is the solubilization ratio, and c is a constant equal to about 0.3 for all hydrocarbons, chlorocarbons and mixtures of these tested to date. This equation shows that as the solubilization ratio (the volume of organic liquid in the microemulsion divided by the volume of surfactant) increases, the interfacial tension decreases. Thus, it is impossible to simultaneously have both high solubilization and high IFT. Any change that is made to the surfactant to increase its IFT in an effort to avoid mobilization will simultaneously decrease its effectiveness in solubilizing the NAPL. Thus, we must accept less efficiency if we want to keep the IFT high. In practice, this means IFTs on the order of 1 mN/m or higher and solubilization ratios on the order of 0.6 or lower. The latter ratio will typically translate into solubilities on the order of 20,000 mg/L or lower. The surfactant requirements under such high IFT conditions are very different than under the ultralow IFT conditions desirable for EOR.

There are many other differences between EOR and SEAR. The environmental acceptability criteria are certainly different. When used in fresh water aquifers, the surfactant should be readily biodegradable and nontoxic at the concentrations used. The reservoir conditions are typically very different. Oil reservoirs are for the most part very deep and have higher temperatures and pressures than aquifers. Although fresh water oil reservoirs do exist and many have been water flooded with fresh water for many years, most of them originally contained high concentrations of electrolytes most often ranging from 1000 to 100,000 mg/L total dissolved solids (TDS). Most of the aquifers that are targets for SEAR are unconfined, whereas all oil reservoirs are confined. This impacts the reservoir engineering more than the surfactant selection, but still along with all of the other differences must be kept in mind when applying EOR technology to SEAR. Table 1 summarizes some of these differences in reservoir conditions. Phenomena of special concern for SEAR include the effect of the natural hydraulic gradient, the solubility of the contaminant in the water, non-equilibrium mass transfer of the contaminant, the adsorption of the contaminant on the soil and biological processes.

Table 1. Comparison Between Typical Oil Reservoir and Aquifer Conditions

	Oil Reservoir	Aquifer
Salinity, mg/L	100 - 300,000	100 - 10,000
Pressure, psi	50 - 15,000	14.7 - 400
Temperature, °F	80 - 250	50 - 70
Depth, ft	100 - 15,000	3 - 900
Permeability, Darcies	0.001 - 10	1 - 100
Hydraulic Gradient, ft/ft	-	0.001 - 0.020
Well Spacing, ft	200 - 5000	30 - 300
Upper Boundary	Confined	Confined or Unconfined
Oil or NAPL Density, g/cc	0.8 - 1.0	0.8 - 1.6

Surfactant Screening.

The criteria for screening surfactants for SEAR are as follows:

* solubilization potential
* phase behavior
* environmental acceptability (toxicity)
* viscosity of surfactant solutions
* coalescence behavior
* cost and availability
* transport characteristics in permeable media
* stability
* sorption characteristics

As in EOR, we take the point of view that the phase behavior is the most fundamental of these requirements for SEAR. This is because the IFT, solubilization, coalescence behavior and other important properties all can be related to phase behavior. We prefer surfactants that when mixed with water and NAPL readily form stable microemulsions rather than liquid crystals, gels or other condensed phases and/or emulsions since these are typically too viscous, show high retention in the rock or

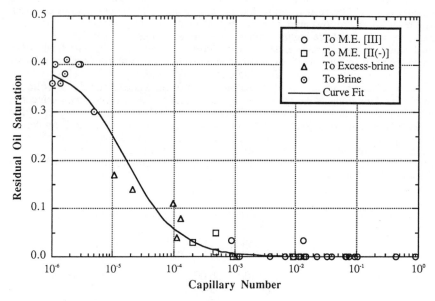

Figure 1.   Measured Capillary Desaturation Curve

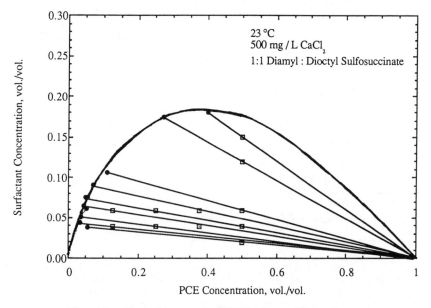

Figure 2.   Phase Diagram for PCE/Surfactant/Water System

soil, are hard to control and predict with models and in general much more risky to use. Historically, EOR surfactants were mixed with alcohols or other co-solvents to prevent or minimize these undesirable characteristics. This is not always necessary at higher temperatures or with surfactants that have a highly branched hydrophobe. Learning how to structure these surfactants to both eliminate the need for co-solvent and have very high solubilization ratios (and thus ultra low IFT) was one of the major accomplishments of the EOR research activity over the past 30 years and there is a rich literature on this subject (*1, 3*).

These microemulsions that we seek when screening surfactants may be Type I (water-rich in equilibrium with excess organic phase), Type III (in equilibrium with both excess water and organic phase) or Type II (oil-rich in equilibrium with excess water). Numerous system variables can then be used to change from one of these forms to another as desired, e.g., the salinity may be used for this purpose. An example of a Type I ternary diagram is shown in Fig. 2. The surfactant used in this case is a sodium sulfosuccinate, which is a food grade sulfonate available commercially. The organic in this illustration is perchloroethylene (PCE), one of the most common contaminants and the subject of our initial SEAR research on DNAPLs (*8-11*). The water in this case contains 500 mg/L of $CaCl_2$ and the temperature is 23°C. One of the advantages of using an anionic surfactant is that electrolytes can readily be used to control the phase behavior. This is illustrated in Fig. 3, which is a plot of the solubilization of the PCE as a function of the $CaCl_2$ concentration. At 500 mg/L, about 80,000 mg/L of PCE have been solubilized in this Type I microemulsion, whereas at the optimal salinity of 1,300 mg/L, 900,000 mg/L of PCE have been solubilized in a Type III microemulsion. The optimal salinity is used here in the traditional sense of the EOR literature (*12*) to mean the point at which the electrolyte concentration in the water results in an equal solubilization of water and oil in the microemulsion, and not in the sense of meaning the best for applications. It is simply a commonly used designation of a certain type of phase behavior. The IFT for this same system is shown in Fig. 4, which shows that the data with PCE follow the same trend as those for decane, as predicted by the Chun Huh equation.

Although a solubilization of 900,000 mg/L may seem high to SEAR researchers, the surfactant used in this study is actually not a very good surfactant compared to the best that have been studied for EOR. The solubilization ratio at optimum is only about 6. The very best EOR surfactants have solubilization ratios on the order of 20, with 10 or higher very common and values as high as 60 observed in some cases. This does not mean that these super surfactants are better for SEAR, but only that they are better at solubilizing organic molecules. What makes these EOR surfactants so good in this sense? The requirement is that both the hydrophilic and lipophilic parts of the molecules have high and balanced affinities for water and oil, respectively. This can be accomplished in part by making the lipophilic part large, i.e., a long hydrocarbon branch is needed. The hydrophilic affinity can be increased by adding ethylene oxide (EO) units to the molecule. This also greatly increases its salt and hardness tolerance. A combination of EO and propylene oxide units (PO) has been found to be a very effective way to engineer surfactant molecules for EOR. The more obvious method of adding a second sulfonate group (sulfate groups are equally effective and just as acceptable at low temperatures where they are chemically stable) usually makes the molecule much too water soluble, i.e., unbalanced. If the hydrocarbon branch is made long and linear, then problems with condensed phases, especially at low temperatures typical of aquifers, is likely. The solution to this is to make the hydrocarbon branched, even though this means giving up some of the very high surfactancy potential. Alternatively, light alcohols such as ethanol or isopropanol can be used to remedy any problems with condensed phases and the use of alcohols with the surfactant has other advantages as well.

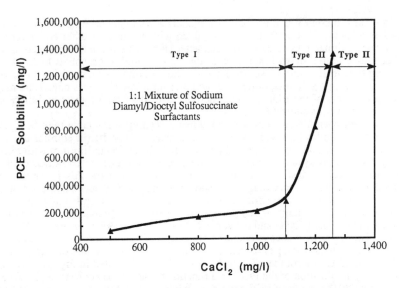

Figure 3.   Solubilization of PCE as a Function of CaCl₂ Concentration

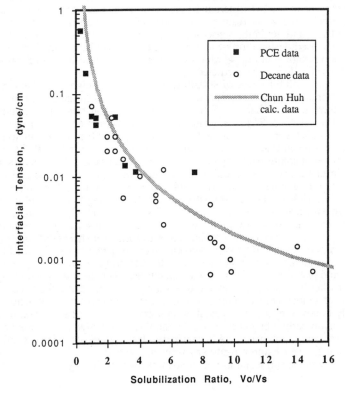

Figure 4.   Comparison of Theoretical and Experimental Interfacial Tension

We very often find it useful to mix different anionic surfactants. A recent series of papers by Baran et al. (*8-10*) has shown how mixtures of sulfosuccinates and ethoxylated sulfates made from branched alcohols called Guerbet alcohols behave with several chlorocarbons, mixtures of chlorocarbons, and mixtures of hydrocarbons and chlorocarbons. An example of the solubilization ratio as a function of the surfactant composition is shown in Fig. 5. Baran et al. have shown that the same mixing rules that were developed earlier for predicting the optimum salinity of hydrocarbon mixtures also work for these mixtures with chlorocarbons. This is very useful information since most waste sites contain complex mixtures of organic contaminants and these mixing rules can be used to predict the behavior of such NAPLs from only the knowledge of the pure component behavior with the surfactant of interest. The remarkable generality of the EACN concept from EOR is illustrated in Fig. 6.

Alternatively, nonionic surfactants may be considered. In this case, the most common hydrophilic moiety is one or more EO groups. The simplest such nonionic is an ethoxylated alcohol. A tremendous amount of research has been done in an attempt to use nonionics for EOR, but so far they have not been found to be competitive with anionics. The general tendency of the ethoxylated alcohols is to form persistent emulsions. Anionics often have the same problem and this tendency increases as the EO number (EON) increases. The best approach is to screen a large number of mixtures under the conditions of interest and observe the rate of coalescence, the fluidity of the phases, the rate of equilibration of the phase volumes and other similar qualitative features and then continue to evaluate only those surfactants that show both reasonably high solubilization ratios and rapid equilibration to microemulsions with low viscosity. This will eliminate most nonionics. Another reason that nonionics have not been competitive with anionics in EOR are the higher adsorption on typical reservoir formation surfaces. In practice almost all surfactant EOR applications have been on relatively clean sandstones, which have a negative surface charge in the pH range of oil reservoirs (nearly always 6.5 to 8.0). Still another reason is that commercially available ethoxylated alcohols have a wide EO distribution that tends to result in separation of the different species, whereas anionics of similar molecular structure, e.g., mixtures of alkyl benzene sulfonates form mixed micelles that show no practical degree of separation in oil reservoirs when used at high concentrations, which is typically 10,000 times higher than the critical micelle concentration (CMC). Mixtures of anionics and nonionics have also met with very limited success in EOR. This does not mean that it is impossible to use them in either EOR or SEAR. Some success has already been achieved using nonionics in SEAR. It simply means these potential problems should be very carefully evaluated and clearly shown not to be problems under the specific conditions of interest with the specific surfactant tested.

Once the phase behavior screening has produced what appears to be an acceptable surfactant for the particular oil or NAPL at the specific temperature and salinity of the reservoir or aquifer, the next significant step in the screening process is to test the displacement characteristics of the surfactant in the permeable medium, a process known as coreflooding in petroleum engineering and column flooding in environmental engineering. Coreflooding is an art with many more complexities than apparent to most researchers. The amount of organic phase displaced and recovered is of course important, but several other observations are equally or even more important since it is relatively easy to find surfactants that will recover almost all of most oils and NAPLs, especially from easy targets such as the typical sandy aquifer formations since they have very large pores, relatively uniform pores and relatively clean pores compared to most oil formations. For example, most oil formations have permeabilities between 0.001 and 10 Darcies (1 Darcy is about 1 μm^2 or 10^{-5} m/s hydraulic conductivity) and clay fractions of 5 to 10% compared to

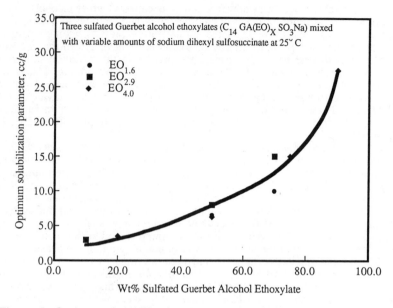

Figure 5. Optimum Solubilization Parameter as a Function of Percentage Guerbet Alcohol

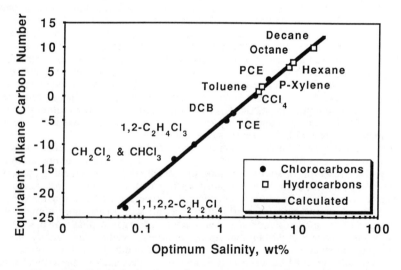

Figure 6. Equivalent Alkane Carbon Number vs. Optimal Salinity

sandy aquifers of 1.0 to 100 Darcy and clay contents of 0 to 5 %. An example column experiment using Ottawa sand is shown in Fig. 7. Most of the PCE in this example was mobilized and almost 100% of it was displaced by 1.5 pore volumes even though the phase behavior was Type I and the IFT 0.09 mN/m, which shows that it is possible to efficiently mobilize and displace DNAPL under these conditions at a much higher IFT than the 0.001 mN/m that is often required to completely displace oil from reservoirs. This example clearly shows that ultralow IFT and Type III behavior are not required to mobilize DNAPL. Careful attention to the capillary number and Bond number under the specific conditions of interest is required to predict whether mobilization of DNAPL will occur rather than the simple idea that it occurs only with Type III microemulsions or below a certain IFT or whatever. Unlike EOR, in SEAR we are also interested in the low concentrations of contaminant in the water after most of the NAPL has been flushed out. Fig. 7 also shows the PCE concentration in the effluent as measured by GC analysis. Even after of 6 pore volumes of 4% surfactant, the concentration is still about 200 mg/L. This is negligible for EOR but not contaminated water. Some of our recent experiments with PCE and TCE demonstrate that we can reduce the concentration of contaminant in the water to at least 10 mg/L after a few pore volumes, but this is still much higher than EPA drinking water standards.

In any case, some of the most important observations include pressure gradient along the core, produced emulsions, retention (as opposed to adsorption, which is usually not a problem with clean formations), relative permeability characteristics, rate dependence of the results and similar indicators of the equilibrium or non-equilibrium behavior, mobility of the fluids, fluid/rock interactions such as cation exchange that are shifted when the surfactant is injected, etc. Assuming mobilization is the objective and a low IFT system is being tested as in EOR, the ideal behavior is to produce almost all of the organic phase before breakthrough of the surfactant followed by clean microemulsion of low viscosity, low retention and low pressure gradient all by 1.2 to 1.5 pore volumes after surfactant is injected into the column. The opposite extreme of no mobilization but only solubilization should similarly show low retention of surfactant, low pressure gradients and clean, low viscosity microemulsions, but of course no clean oil bank is produced first and the recovery of the NAPL requires at least several pore volumes in this approach, often 10 to 20 pore volumes. The single most common mistake made by those screening surfactants for EOR has been to not measure pressure drop or not measure it accurately. The low pressure drops typical of many properly designed coreflood experiments can be difficult to measure accurately. This is more of a problem with typical column experiments than typical oil corefloods since the permeability of the soil from the aquifer is likely to be higher and thus the pressure gradient lower unless the flow rate is increased by many fold, which is a mistake since then non-equilibrium phenomena that are not scaled to the aquifer become significant. In other cases, high pressure gradients along the core are measured and then ignored. Even in the early screening process, these measurements should be done and taken into account since they are the most sensitive indication of problems with emulsions and other undesirable or unexplained phenomena.

Mobility Control.

Another consideration in screening surfactants for surfactant EOR is mobility control. The polymers that are typically used for mobility control in EOR mostly effect the viscosity of the solutions as desired, but also do have some effect on the phase behavior, adsorption and other properties of the micellar/polymer mixtures. The two most commonly used polymers are hydrolyzed polyacrylamide (HPAM) and xanthan gum, but there are several other good choices. The molecular weight of these polymers is in the range of 1 to 20 million and both of these polymers have the negatively charged carboxylate group as an active group affecting their water solubility.   These polymers will typically shift the phase boundaries of micellar systems slightly and this effect is not important, but sometimes a much more important effect occurs as a result of either microgel formation induced by the polymer or the water-rich phase sometimes separates into one phase containing most of the polymer and one phase containing most of the surfactant, which is undesirable and important. Therefore, even at the screening stage, some phase behavior studies including polymer should be made. Some of the classical remedies to such problems when they occur are to lower the salinity, add co-solvent or a small amount of hydrocarbon to the microemulsion, lower the surfactant concentration, or change surfactant or polymer. Since polymer is such a critical element to successful EOR field applications, and may very well be just as important ultimately to SEAR applications, experience with EOR polymers is briefly described next.

The most important benefit of EOR polymers is the improvement of sweep efficiency. When the injected fluids are thickened to about 10 to 20 times the viscosity of the reservoir brine at reservoir temperature, much less bypassing of oil occurs in the oil reservoir and much less injected fluid is required to displace the oil that is contacted. The favorable attenuation of vertical heterogeneity effects is especially important. The importance of the benefits of using polymer can be emphasized by pointing out that there has never been a technically successful surfactant EOR field project that did not use polymer. In addition, any tendency for the injected surfactant solution to finger through the oil bank can be eliminated with the appropriate amount of polymer. Since the surfactant increases the relative permeability of the water about 10 to 20 fold in most cases, a compensating increase in the water viscosity is required to maintain viscous stability. This typically requires only about 0.1 % polymer, and the cost of the polymer is about the same per pound as surfactant, so the cost of polymer is less than for surfactant. In SEAR applications, the relative permeability behavior is typically more favorable than in sandstone oil formations, so about 5 cp water viscosity is likely to be adequate in most cases, which would require only about 0.05% polymer, two orders of magnitude less than the surfactant concentration. The viscosity of the oil matters very little with respect to mobility control in the range of 1 to 10 cp typical of light crude oils. This counter intuitive result is because the behavior is dominated by the relative permeability, which needs to be measured accurately as a result.

HPAM also effects the mobility of the fluids by permeability reduction. However, this complex effect diminishes as the permeability increases and is not likely to be very great above about 10 Darcy. Furthermore, xanthan gum is the more likely choice for SEAR since it is an FDA approved food additive and presumably there would not be any concerns about using it in aquifers along with surfactant. Polymer has also been found to decrease the surfactant retention under some conditions. One mechanism for this is competitive adsorption. However, the more important mechanism is likely to be better microscopic sweep of the pores as the displacing fluid viscosity increases. Under conditions where surfactant retention decreases when polymer is added, the polymer can actually decrease the cost of the EOR operation.

These EOR polymers show complex non-Newtonian rheology. There is a rich literature on this subject (13) and no attempt will be made to address it here except to point out that the apparent viscosity in the permeable medium is somewhat less than the bulk viscosity measured at low shear rate in couette viscometers or similar rheometers. This would need to be taken into account in SEAR applications also. There are other complexities such as elasticity for some polymers, although this happens to not be very important with xanthan gum. Other complexities that must be dealt with and have been extensively investigated during the past 30 years and are in the petroleum literature are polymer adsorption, inaccessible pore volume, shear degradation, thermal stability, biodegradability, electrolyte compatibilities and the like. Fortunately, none of these except biodegradation should be of any significance with respect to SEAR applications because of the low temperature, low salinity, high permeability, low surface area conditions of aquifers compared to oil reservoirs.

Other methods of mobility control besides polymer have been used in EOR and these deserve at least brief mention. When a dilute aqueous surfactant solution and gas are alternately injected in a permeable medium, a phenomena loosely referred to as foaming occurs under some conditions. Although this should not be visualized as an actual foam such as can easily be made with bulk fluids containing surfactants as is common in commercial household products, nevertheless it is called "foam" in the petroleum literature (*14*). Water containing the surfactant alternates with gas in the pores to form lamellae and above a certain velocity, surfactant concentration and so forth this effect increases the resistance to flow appreciably. During steady state coreflooding tests, this then looks like an apparent viscosity of anywhere from a few times water to hundreds of times water viscosity. Thus, foam could be used as a mobility control alternative to polymer in SEAR if carefully tailored to the aquifer conditions and requirements and would have the same benefits of increased sweep efficiency as a polymer. The gas in this case would likely be air and some of the same surfactants used in EOR make good foams under conditions of interest. Emulsions have also been evaluated for EOR, but these seem both less likely to work and less applicable to SEAR than either polymer or foam.

Characterization.

Characterization of both the formation and fluids is extremely important in EOR and it seems likely will be equally important in SEAR as well as other remediation technologies. There have been more EOR failures because of poor reservoir characterization than for all other reasons combined. This observation also applies to other EOR methods such as the use of solvents, microbes and heat. The geology of oil reservoirs is typically very complex and must be taken into account on the scale of at least 1000s of feet for commercial EOR projects. Although some characterization research was done decades ago, it was not until the late 1970s that the industry finally recognized its critical importance to EOR and began large scale research on better characterization methods. This subject is far too large to even summarize here. Although no modern textbooks are known to us, the reader is referred to an extensive literature by the Society of Petroleum Engineers (SPE) and to authors such as Lake. Of course, there are many geological features common to aquifers and oil reservoirs and the ground water literature in some areas such as the application of geostatistics led the petroleum literature in many respects. There are also differences that need to be taken into account, e.g., the scale of remediation processes to date at least has been smaller than that of EOR processes by about one to two orders of magnitude. On the other hand, the need for small scale characterization of aquifers may very well be greater than for oil reservoirs because of the need to minimize the risk with better control of the process. Characterization is of limited value if it is not effectively used in modeling of the fluid flow processes in the reservoir. Modeling will be briefly addressed in the next section.

In addition to geology, the characterization of the media (rock or soil) is important and both applications involve the need to know characteristics such as capillary pressure and relative permeability. This is known as petrophysics in the petroleum literature and also includes the electrical properties and others needed for electric and magnetic logging purposes. By far the most common methods for obtaining these properties are special core analysis and a wide variety of logs. The relative permeability properties are especially important and should always be

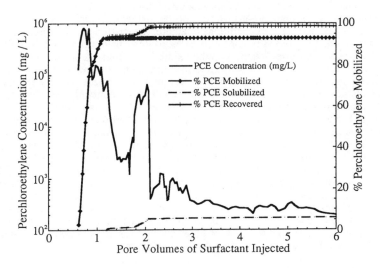

Figure 7. Example of Column Experiment: PCE Recovery and Effluent
Concentration

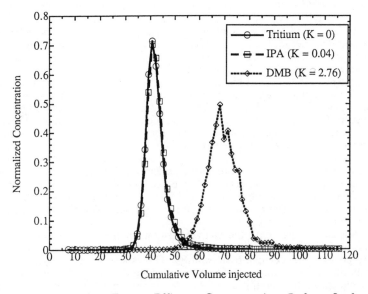

Figure 8. Partitioning Tracer Effluent Concentration Before Surfactant
Remediation

measured. Although complexities such as hysteresis have been known for decades and papers on both measurements and theory are in the petroleum literature, it is perhaps not quite as important as in contaminant transport and remediation. This is because the spill event is mostly a first drainage event and the ground water flow is mostly an imbibition event and both are important in modeling these processes for various purposes.

Fluid characterization depends very strongly on the EOR method so generalizations are difficult, but for surfactant EOR the crude can often be treated as just one big pseudo-component (except for large amounts of the very lightest hydrocarbons such as methane if these are dissolved in the crude at high pressure) with certain effective properties with respect to the surfactant behavior. This may not be the case with complex waste mixtures since they sometimes contain a much wider variety of molecules than crude oil, so more attention to fluid composition and behavior will likely be needed in some SEAR applications.

Another characterization method common to both oil field applications and aquifers is the use of tracers. Tracers have been found to be very useful for detecting thief zones in oil reservoirs, determining geological continuity over large distances, measuring swept volume of the reservoir, evaluating EOR potential and performance and many more purposes, but especially for assessing heterogeneity of reservoirs. Recent research in this area has been done and can be found in DOE reports by Pope et al. (*15*). In addition, if partitioning tracers are injected, then an estimate of the residual oil saturation can be easily and usefully made (*16*). Jin et al. (*17*) have recently proposed and extended this idea for detecting and characterizing NAPL residuals in soils. Since knowing precisely how much NAPL is present and where it is in the subsurface is critically important and often poorly known by any other method or data from most waste sites, this oilfield technology would seem to be not only applicable, but crucially significant to SEAR and other applications. This technology is applicable to both saturated and unsaturated zones. Liquid tracers that partition between the injected water and the NAPL are appropriate in the saturated case and gas tracers that partition between the injected gas (air) and NAPL are appropriate for the unsaturated case. The use of single well backflow tracer techniques (which rely on reactive tracers such as esthers as developed by Deans, Sheely and others) has been successfully used in oil wells more than 100 times and may also have some applicability in aquifers (see *18* for a recent innovation and literature), but the interwell version is more likely to be the useful approach.

Fig. 8 shows an example of the separation of two alcohols when injected as dilute aqueous solutions into a column containing PCE at a residual saturation of about 20%. The lighter alcohol (isopropanol) has negligible partitioning into the PCE and transports at the same rate as the water itself (as verified by comparison with tritiated water). The heavier alcohol (1,2 dimethyl 2-butanol) partitions into the PCE and lags the IPA as a result. From this clear chromatographic signal, one can easily calculate the residual oil saturation. Under laboratory conditions, these measurements agree with independent mass balance measurements to within about 0.02 saturation units. The accuracy of field measurements is harder to estimate since all other methods are more uncertain under oil field conditions, but it appears to be possible (*16*) to estimate residual oil saturation in oil reservoirs to within about 0.03 saturation units over distances of more than 1000 feet in highly heterogeneous reservoirs. These same or similar water tracers can be repeatedly applied as a performance assessment tool. Fig. 9 shows an example of this. These same alcohols did not separate when injected after the surfactant remediation, which from independent measurements was known to have removed at least 99.5% of the PCE from the column. Similar measurements have been made with other contaminants such as trichloroethylene (TCE) and jet fuel (JP4) with similar or better success in our laboratory and most recently with the use of partitioning gas tracers in unsaturated soils containing TCE (*19*).

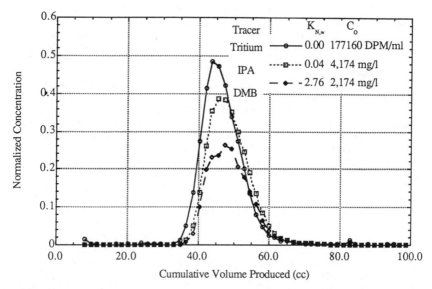

Figure 9. Partitioning Tracer Effluent Concentration After Surfactant Remediation

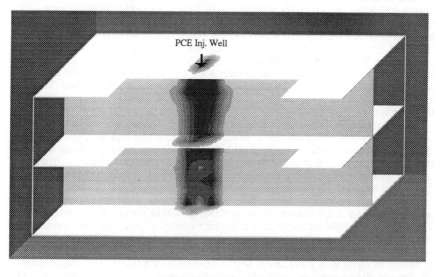

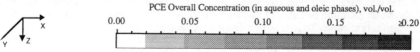

Figure 10. PCE Concentration Contours During Contamination Event After 30 Days

There are of course differences between oil field tracer applications and those to SEAR and these differences need to be taken into account. As with the surfactant itself, environmental acceptability is more stringent. Thus, commonly used radioactive tracers such as tritium that make wonderful oil field tracers are not likely to be acceptable in aquifers even at very low levels. There are a variety of more subtle differences. The distribution of the NAPL is more uneven and complex in typical subsurface waste sites than residual oil saturation is in typical oil reservoirs. Most waste sites are unconfined on top. Most waste sites contain complex organic mixtures and lower average residual oil saturations and some of the contaminant may be adsorbed on natural organic carbon or the soil itself and some will be in the water phase. Well distances, completions and injection/production methods are different. Shallow soils are often aerobic whereas oil reservoirs are anaerobic. Thus, there are both differences that need to be taken into account in applying tracers and potential innovations that should be researched.

Modeling.

Modeling of fluid flow in permeable media has a long and distinguished history in both the ground water and petroleum literature. Its importance for design, optimization, scaleup and performance prediction and assessment has clearly been widely recognized by both communities. Abriola (20) summarizes the literature very well up to 1989 and compares the two approaches. Perhaps the most significant difference in terms of both need and approach is that until recently most of the ground water modeling was of single phase flow problems and conditions, whereas out of necessity most of the petroleum reservoir modeling research has been multiphase flow problems and conditions. Since multiphase flow is needed to simulate both the initial conditions and the displacement that occurs for SEAR applications, the multiphase flow modeling developments in the petroleum literature are highly pertinent, although once again there are significant differences in conditions between EOR and SEAR that must be taken into account.

It was for this reason that The University of Texas (UT) and The University of Michigan jointly developed and applied a modified version of a chemical reservoir simulator known as UTCHEM that was developed at UT during the 1980's for EOR applications (21). The results of this effort can be found in Brown et al. (22) and Delshad et al. (23). Some of the modifications that had to be made to UTCHEM were (1) the partitioning of the contaminant from the NAPL to the water including non-equilibrium mass transfer; (2) open boundaries on the sides and top of the aquifer; (3) adsorption of the contaminant; (4) first drainage capillary and relative permeability functions (we also added new models that are commonly used in ground water modeling but not petroleum modeling); and (5) the addition of an air phase with tracers. We are currently adding other features such as biological reactions and additional geochemistry. Even with all of these changes, the previous modeling development and experience with surfactant EOR proved to be very valuable.

An example of the simulation of a PCE spill in a saturated soil similar to that at the Borden site is shown in Fig. 10. This simulation illustrates the combined effects of capillarity, gravity, heterogeneity and the hydraulic gradient on the transport and capillary trapping of the PCE as it percolates downward through the soil. During the second phase of the simulation, water flowed past the trapped contaminant for 60 days under natural gradient conditions. Some PCE dissolves into the water and is transported in the water by both advection and dispersion. Next surfactant remediation was simulated. An example of the distribution of the surfactant in the aquifer after 60 days of surfactant injection is shown in Fig. 11. Even with four injectors in this balanced five spot well pattern, there has been a significant distortion of the surfactant caused by the natural hydraulic gradient. This effect can be minimized by either injecting at higher rates or using a line drive or perhaps by other means. This is one of many examples of the useful insight that can be obtained from such simulations.

Field Testing.

The final step in evaluating an EOR technology is conducting one or more field pilot tests. Hundreds of these have been done by the oil industry worldwide during the past 30 years and it would require an entire book to summarize this experience. Because of both the geological complexities and process complexities, these pilot tests are essential in developing, evaluating and demonstrating EOR technology. Similar testing of the SEAR technology is in progress, although to date the scale of the SEAR tests has been much smaller and more confined, e.g., a test cell was used at Borden which was only 3 m on a side (*24*). The most important lesson learned from all of these EOR pilots is that they not only need to be done, but that they need to be done using the best available means of design (this means accurate models among other things), control and measurements both before the test (this means careful characterization with logs, cores, tracers, pressure tests, laboratory corefloods, etc.) and during the test (this means using observation wells, complete compositional analysis of produced fluids, pressure monitoring, etc.). We are particularly struck by the compelling need to use the partitioning tracer concept appropriately modified and applied to aquifers to estimate the NAPL saturation (or some other equivalent technology, but we know of none) since the failure to accurately estimate residual oil saturation in the reservoir before the EOR field pilot or commercial project caused more grief and wasted cost than perhaps any other single mode of failure and both the need to know this accurately and the uncertainty of it appear to be greater in the contaminated soil case than in EOR. Any attempts to save money or cut corners almost always ends up costing more money because another expensive pilot will have to be done later, excessive efforts will be expended trying to interpret incomplete and inadequate pilot data, large expenditures will be made on post pilot coring and the like. More problematic than the additional costs of a cheap pilot is the delay in the development of the technology, the confusion and controversy in its interpretation and the impact that this has on ongoing process research, future plans, economic projections and so forth. Those companies that made a sustained effort to test surfactant EOR with well designed and operated pilots on a systematic basis over a period of time eventually learned how to use this complex technology successfully.

Of course these EOR pilots are much more expensive than SEAR pilots because the wells are much deeper, the scale is much larger and the conditions much harsher, so differences in approach are appropriate. On the other hand, the requirements for safe and environmentally acceptable pilots and the criteria for success is more severe for remediation of soils than it is for EOR. The demands on our SEAR models will be significantly greater as a result. Until the 1980s (recall this is when the industry finally achieved technical success on a large scale with surfactant EOR) the oil industry often did not model the pilot before it was conducted and sometimes not even afterwards. This is a fatal mistake. There can never be a rational excuse for conducting a field test in the subsurface without careful and systematic modeling. When used with good engineering judgment, a good model will yield valuable insight, help identify critical data needs, minimize risk of failure, save time and money, and in many other ways pay off. The fact that the model may not be completely validated increases the incentive to use it to predict and interpret the pilot so that progress can be made in this validation process. Of course, modeling is not the only tool of the reservoir engineer and should not be used in a blind or isolated sense.

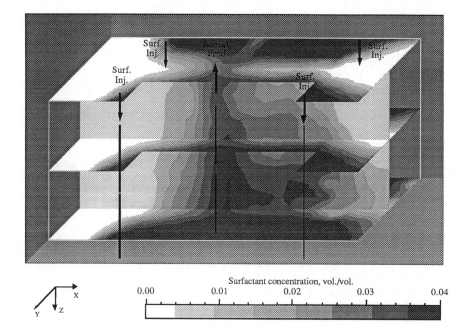

Surfactant concentration, vol./vol.

0.00   0.01   0.02   0.03   0.04

X
Y   Z

Figure 11. Surfactant Concentration Contours After 60 Days of Surfactant Injection (Day 150)

Another characteristic of companies with successful track records in EOR is the use of a fully integrated team of reservoir engineers, geologists, geophysicists, chemists, petrophysicists, chemical engineers and others as appropriate. Although the list might be somewhat different in the case of SEAR testing, e.g., it might also include environmental engineers, contaminant hydrogeologists and microbiologists, the concept is the same. Although this is a far from complete summary of the lessons learned from EOR, this will be the final point made in this brief overview, and perhaps the most important one since without the appropriate expertise and experience of such a team, none of the other lessons are likely to be fully understood and heeded. The reader is encouraged to study the EOR literature (which space allowed us to reference only a tiny fraction) for more lessons and more detailed information on these lessons.

Acknowledgments.

We would like to thank C.L. Brown for help in preparing this manuscript and M. Jin and V. Dwarakanath for the experiments and simulations shown as illustrations. We would also like to acknowledge the many helpful discussions that have occurred during the past three years of collaboration on SEAR between the authors and L.M. Abriola, K.D. Pennell, R.E. Jackson (who especially played a key role in the development of the partitioning tracer ideas), D.C. McKinney and numerous others and the other research staff involved in one or more of the related research projects in Petroleum Engineering, especially K. Sepehrnoori, M. Delshad and B. Rouse.

References.

1. Lake, L.W. *Enhanced Oil Recovery*, Prentice Hall, Englewood Cliffs, New Jersey, **1989**.
2. Pope, G.A.; Bavière, M. *Basic Concepts in Enhanced Oil Recovery Processes*, Critical Reports on Applied Chemistry, Volume 33, M. Bavière (ed.), Elsevier Science Publishing, **1991**.
3. Bourrel, M.; Schechter, R.S. *Microemulsions and Related Systems*, Marcel Dekker, Inc., **1988**, 30, Chpts. 4-7.
4. Delshad, M. *Trapping of Micellar Fluids in Berea Sandstone*, Ph.D. dissertation, The University of Texas, Austin, **1990**.
5. Morrow, N.R.; Songkran, B. *Surface Phenomenon in Enhanced Oil Recovery*, Plenum, New York, **1981**.
6. Huh, C. *J. Colloid and Interface Science*, **1979**, *71*, 408-426.
7. Sunwoo, C.; Wade, W.H. *J. Dispersion Science Technology*, **1992**, *13*, 491.
8. Baran, Jr., J.R.; Pope, G.A.; Wade, W.H.; Weerasooriya, V. *Langmuir*, **1994**, *10*, 1146-1150.
9. Baran, Jr., J.R.; Pope, G.A.; Wade, W.H.; Weerasooriya, V.; Yapa, A. *Environmental Science and Technology*, **1994**, *28*, 1361-1366.
10. Baran, Jr., J.R.; Pope, G.A.; Wade, W.H.; Weerasooriya, V.; Yapa, A. *J. Colloid and Interface Science*, **1994**, *168*, 67-72.
11. Pennell, K.D.; Jin, M.; Abriola, L. M.; Pope, G. A. *J. of Contaminant Hydrology*, **1994**, *16*, 35-53.
12. Reed, R.L.; Healy, R.N. *Improved Oil Recovery by Surfactant and Polymer Flooding*, Academic Press, New York, **1977**, 383-437.
13. Sorbie, K.S. *Polymer-Improved Oil Recovery*, CRC Press, Inc., Boca Raton, FL, **1991**.
14. Hirasaki, G.J. *J. Petroleum Technology*, **1989**, May, 449-456.
15. Pope, G.A.; Sepehrnoori, K.; Delshad, M. *U.S. Department of Energy*, U.S. DOE Contract DE-AC22-90BC14653, **1994**.
16. Allison, S.B.; Pope, G.A.; Sepehrnoori, K. *J. of Petroleum Science and Engineering*, **1991**, *5*(2).
17. Jin, M.; Delshad, M.; McKinney, D.C.; Pope, G.A.; Sepehrnoori, K.; Tilburg, C. American Institute of Hydrology Conference, **1994**, 131-159.
18. Ferreira, L.E.A.; Descant, F.J.; Delshad, M.; Pope, G.A.; Sepehrnoori, K. presented at *SPE/DOE Eighth Symposium on EOR*, Tulsa, OK, **1992**.
19. Pope, G.A.; McKinney, D.C.; Whitley, Jr., G.A.; Mariner, P.E. to be presented at the *1995 International Symposium on In Situ and On Site Bioreclamation*.
20. Abriola, L.M. *Environmental Health Perspectives*, **1989**, *83*, 117-143.
21. Datta Gupta, A.; Pope, G.A.; Sepehrnoori, K.; Thrasher, R.L. *SPE Reservoir Engineering*, **1986**, 622-632.
22. Brown, C.L.; Pope, G.A.; Abriola, L.M.; Sepehrnoori, K. *Water Resources Research*, **1994**, *30*(11), 2959-2977.
23. Delshad, M.; Pope, G.A.; Sepehrnoori, K. submitted to *Journal of Contaminant Hydrology* **1995**.
24. Kueper, B.H.; Redman, D.; Starr, R.C.; Reitsma, S.; Mah, M. *Ground Water*, **1993**, *31*, 756-766.

RECEIVED February 10, 1995

Chapter 12

Evaluating Effectiveness of In Situ Soil Flushing with Surfactants

Katherine A. Bourbonais, Geoffrey C. Compeau[1], and Lee K. MacClellan

AGI Technologies, 300 120th Avenue, Northeast, Bellevue, WA 98005

Laboratory treatability studies were conducted to evaluate the suitability of surfactants for in situ flushing of diesel and motor oil petroleum hydrocarbons and other organic contaminants from gravelly soil. The objective of the studies was to determine a flushing regime for soil at a U.S. Environmental Protection Agency (EPA) National Priorities List (NPL) site. Target contaminants of concern include volatile organic hydrocarbons (VOC), polycyclic aromatic hydrocarbons (PAH), and various metals. Twenty-eight nonionic, anionic, and mixed anionic/nonionic surfactant formulations were tested. Screening tests performed with aqueous surfactant solutions (0.01 to 1.0 percent, v/v) evaluated soil colloid dispersion, petroleum hydrocarbon dispersion, solubilization, and surfactant detergency. The four most promising candidates were further evaluated in soil column tests and leachate characterization studies. Results indicated greater than 90 percent reduction of soil total petroleum hydrocarbons (TPH), up to 98 percent reduction of VOCs, and 49 to 99 percent reduction of various 3- through 6-ring PAHs. However, this study also identified drawbacks to in situ surfactant flushing at this site, including mobilization of soil fines, difficulty in removing surfactant residue from soil, potential difficulty and expense in treating recovered leachate, potential surfactant instability in the subsurface, and microbial degradation potentially resulting in anaerobic conditions and impacting subsurface characteristics.

[1]Current address: Enviros, Inc., 25 Central Way, Kirkland, WA 98033

Petroleum hydrocarbons (commonly analyzed as TPH) released from a surface or buried container permeate downward and laterally through the soil. In the absence of an impermeable barrier, the limit of downward movement is typically the upper surface of an aquifer. As petroleum hydrocarbons reach the aquifer, they begin to spread horizontally and form a lens on the water surface.

TPH that permeates the pore structure of the aquifer boundary will be dispersed into droplets, or ganglia, whose size range is determined by interfacial tension and capillary dimensions of the pores. Some fuel components adsorb onto surfaces within the capillary fringe and in the overlying unsaturated soil. A portion of the TPH can typically be recovered as product via an extraction well or interception trench; however, a portion will remain entrapped in the pore structure at the aquifer boundary. Recovery of this material requires mobilization, which can be accomplished by several means, including reducing the surficial forces responsible for droplet retention.

Surfactants can be used to reduce interfacial tension and assist in solubilizing hydrophobic hydrocarbon constituents. Using surfactants for flushing contaminated soils is the logical extension of secondary oil recovery techniques applied in the petroleum industry. Here, historical precedent exists for using surfactants to lower interfacial tension between hydrocarbon phases and water, thus enhancing extraction of oil from porous media. A brief overview of some of the research which has been done to adapt surfactant technology to environmental remediation applications follows.

Nash (1) conducted laboratory and field studies to evaluate the effectiveness of surfactants in removing TPH and chlorinated hydrocarbons from soil. Laboratory results showed promise, with removal efficiencies of 88 percent for topped Murban crude oil and 90 percent for polychlorinated biphenyl (PCB) contamination; however, a subsequent field trial conducted at an Air Force fire training area did not demonstrate significant removal. Factors that may have affected surfactant performance in the field are losses through adsorption, precipitation, and phase-trapping; and chromatographic separation of the surfactant into individually ineffective components. The authors observed that downward transport of hydrocarbons appeared to be taking place during the 7 days that soil flushing was conducted.

Vignon and Rubin (2) assessed a variety of alkylphenolethoxylated and alkylethoxylated surfactants for activity and optimal dose required to solubilize sorbed anthracene and biphenyl. They concluded dosages greater than 0.1 percent w/v were needed to effect significant desorption. Rickabaugh, et al. (3) examined surfactant-enhanced soil scrubbing as a technique for aqueous separation of chlorinated hydrocarbon contaminants from soil. Target compound removals ranged from 28 to 59 percent. Batch and column results showed blends performed better than any single surfactant alone, and 2.0 percent solutions were more effective than either 1.0 or 0.5 percent.

Abdul, et al. (4,5) investigated the suitability of 10 commercially available surfactants for removal of petroleum constituents from sandy soils, and field-tested an in situ surfactant-flushing process to decontaminate soils containing

PCBs and oil. Surfactant selection for the field demonstration incorporated results of screening the 10 candidate surfactants for minimum surface tension of an aqueous solution, extent of soil and oil dispersion, extent of oil solubilization, and effectiveness at washing oil from the soil in batch tests.

The field test demonstrated removal of 10.5 and 10.7 percent, respectively, of the PCBs and oils after flushing with surfactant for 70 days followed by 30 days rinsing. They concluded that permeation and washing of the target zone could be accomplished without significant lateral spread of surfactant or leachate if flushing rate is carefully controlled. Furthermore, complete leachate recovery from the aquifer could be achieved with minimal extraction of excess groundwater by controlling recovery rate. However, the study test plot was surrounded by a cement-clay containment wall.

Subsurface contamination by residual surfactant solution is typically not a concern during enhanced crude oil recovery operations. In environmental restoration applications, however, the surfactant itself must not cause further environmental impacts. Additionally, regulatory approval for injection of surfactants into the subsurface has to be obtained. Recently, some researchers have focused on the use of edible surfactants in an effort to enhance the likelihood of obtaining regulatory approval.

Shiau, Sabatini, and Harwell (6) have conducted research on surfactants with U.S. Food and Drug Administration direct food additive status, and found edible surfactants capable of middle phase microemulsion formation and micellar solubilization of perchloroethylene (PCE), trichloroethylene (TCE), and trans 1,2-dichloroethylene (DCE). Micellar partitioning coefficients were comparable to values reported for other types of surfactants.

Site Background

The Record of Decision issued for the subject NPL site established requirements for remedial design and remedial action, including in situ soil flushing. Successful surfactant performance in flushing contaminants was predicted based on the previously referenced studies. The specific objective of this treatability study was to identify a flushing solution and regime most likely to achieve remediation performance goals, which were based on protection of groundwater quality.

Study results were expected to be used as the basis to select an appropriate flushing solution, determine optimal surfactant concentration ranges, determine the approximate effect of soil flushing on soil target contaminant concentrations, and provide general indications of leachate quality and contaminant concentration following implementation of soil flushing. The site is a large diesel engine maintenance and repair facility. Site operations historically included use of various hydrocarbons and cleaning agents, including fuels, lubricants, detergents, and halogenated and nonhalogenated hydrocarbon solvents. The source of contamination is an unlined pit used to dispose of sludge generated by an on-site wastewater treatment plant. The pit covers approximately 1 acre.

Geology and Hydrogeology. Silty alluvial soil mantles the land surface around the pit. Parts of the pit are underlain by this silt, which appears to have been physically and chemically modified, presumably by leachate originating in the overlying sludge. In other areas, sludge rests directly on the upper layer of a gravel deposit, which appears very darkly stained. In contrast, gravel encountered below the silt deposit is generally unstained.

The gravel deposit underlies the alluvial soil and extends approximately 35 feet below ground surface. It consists of an unsorted and poorly stratified mixture of gravel, cobbles, and boulders, with discontinuous layers of sand, silt, and clay. Pore spaces up to 1/2 inch in diameter have been observed, suggesting exceptionally high permeability. The lower 2 to 13.5 feet are saturated and comprise an aquifer. An aquitard underlies the gravel and hydraulically separates the aquifer from deeper water-bearing zones.

Remediation. Remedial actions to be performed at this site include excavation of the sludge and underlying soil to the practical limits of excavation. Remaining contaminated soils were to be treated by in situ flushing. Based on field investigations, an estimated 100,000 cubic yards of underlying soil, primarily gravel, would require treatment. Representative samples were collected from this region for treatability testing.

Sample Collection and Baseline Chemical Characterization

Soil. Study soil was collected at three locations known to be affected by contaminants which had migrated from the source. The soil was screened (1/2-inch U.S. sieve) and field composited. Samples were collected for baseline chemical analyses, including VOCs as measured by EPA Method 8010/8020 (7), PAHs by EPA Method 8310 (7), pH by EPA Method 9045 (7), TPH by EPA Methods 418.1 Modified (8) and 8015 Modified (7), and 14 metals by EPA 6010/7000 series methods (7). Samples were analyzed in triplicate to ensure accuracy and precision.

TPH was measured by two methods: infrared spectroscopy (TPH/IR, EPA Method 418.1 Modified) and gas chromatography with flame ionization detection (TPH/GC, EPA Method 8015 Modified). TPH/IR methods are non-specific and quantitate all freon-extractable nonpolar hydrocarbons present; TPH/GC analyses identify the boiling point profile of the extracted hydrocarbons and quantitate concentrations of different fuel types.

Chemical characterization results (Table I) indicated the soil has a low moisture content, slightly alkaline pH, and high TPH concentration. The TPH/GC profile indicated a fairly unweathered material with diesel-, motor oil-, and heavier oil-range components. Chlorinated aliphatic and aromatic VOCs were detected at concentrations ranging from 1.4 to 21 milligrams/kilogram (mg/kg). All 15 PAHs analyzed were detected, at concentrations ranging from 0.045 to 50 mg/kg. For ease of presentation, PAH data are grouped into two classes: 2- to 3-ring compounds and 4- to 6-ring compounds, based on structure and attendant solubility, mobility, toxicity, and persistence characteristics, and regulatory concerns. Baseline concentrations were 68.6 mg/kg of 2- to 3-ring compounds, and 8.2 mg/kg of 4- to 6-ring compounds. Listed results on Table I and all subsequent tables represent the mean of two or three replicate analyses.

All metals analyzed were detected. Comparison of results to background subsurface soil metals data collected during previous investigations revealed only two metals were present at concentrations significantly exceeding background levels. Metals concentrations were tracked throughout the study in soil and leachate samples. Study results, however, did not indicate significant effects of surfactant treatment on metals. Therefore, metals mobility was not considered to impact flushing solution recommendations, and metals results are not presented in this paper.

Table I. Soil Baseline Chemical Characterization Results

Analyte	Concentration	Analyte	Concentration
pH	8.2	Polycyclic Aromatic Hydrocarbons (mg/kg)	
Moisture Content	6.5 %	Total, 2 to 3 ring	68.6
TPH (mg/kg)	25,000	Naphthalene	2.9
		Acenaphthene	0.93
Volatile Organic Compounds (mg/kg)		Fluorene	2.3
1,2-Dichlorobenzene	14	Phenanthrene	12
1,4-Dichlorobenzene	1.8	Anthracene	0.50
cis-1,2-Dichloroethene	1.4	Fluoranthene	50
Ethylbenzene	4.3		
Toluene	21	Total, 4 to 6 ring	8.2
Total Xylenes	14	Pyrene	1.4
		Benzo(a)anthracene	3.4
		Chrysene	1.1
		Benzo(b)fluoranthene	0.68
		Benzo(k)fluoranthene	0.39
		Benzo(a)pyrene	0.48
		Dibenzo(a,h)anthracene	0.045
		Benzo(g,h,i)perylene	0.34
		Indeno(1,2,3-c,d)perylene	0.39

Water. Site tap water was collected in bulk at the same time as soil samples. To enhance applicability of laboratory results to field conditions, this tap water was used as the water supply for all treatability testing unless specified otherwise. Aliquots were removed at the time of collection and analyzed for the same parameters listed above. Five metals and four halogenated byproducts of water disinfection were detected at low concentrations. The presence of these analytes was not considered likely to have significant impact on test results and tap water was used throughout the study without further testing.

Surfactant Selection

Prior to initiating laboratory studies, surfactants were selected for screening based on a literature search and demonstrated performance. Selection criteria included toxicity and biodegradability, soil flushing performance, solubility in oil and water, detergency, soil colloid dispersion, hard water tolerance, treatment requirements of spent surfactant solutions, and cost. The following general chemical classes of commercially available surfactants were surveyed: thioethoxylates, alcohol ethoxylates and alkoxylates, amine-substituted alcohols, polyoxy carboxylates, polyoxyalkylated phosphate esters, polyalkylene glycol ether, sulfonates, alkylbenzene sulfonates, and alkyl phenoxyether sulfates.

Twenty-eight surfactants (Table II) were selected for screening in tests designed to evaluate relative ability to emulsify motor oil and hydrocarbons adsorbed onto site soil. Anionic, nonionic, and an anionic/nonionic blended product were tested. Other surfactant groups, such as amphoteric or cationic surfactants, were not included because of their expected strong complexation with soil minerals (9).

Laboratory Studies

Screening. Phase I evaluated interactions between surfactants, soil, hydrocarbons, and soil/hydrocarbon mixtures. Screening tests were conducted with aqueous solutions of 0.01 to 1.0 percent surfactant and 99 to 99.99 percent tap water except as noted. These tests evaluated the potential for formation of colloid-size particles of soil, the ability of surfactant solutions to emulsify motor oil and solubilize hydrocarbons from site soil, and interaction between tap water constituents and surfactants in the absence of soil. Phase II quantitatively evaluated surfactant detergency; the relative effectiveness of surfactants in removing TPH from site soil and keeping it suspended in an aqueous medium; and optimization of rinse procedures and surfactant concentrations.

Phase I screening tests were performed by dispensing soil or motor oil and surfactant solutions into 40 milliliter (ml) glass vials. Vials were mixed on a gyrotary shaker for 30 minutes at 150 revolutions per minute (rpm), and turbidity measured immediately. Increases in turbidity were expected to be due at least in part to solubilized hydrocarbons, thus roughly indicating surfactant performance under test conditions.

Turbidity results were not always found to be meaningful indicators of surfactant effectiveness. In some cases, surfactant solutions themselves were very turbid and turbidity decreased in the presence of hydrocarbons. In other cases, surfactant/hydrocarbon mixtures created an oily film or opaque streaks

Table II. Surfactants Tested

Manufacturer and Surfactant	Nionic (N) or Anionic (A)	CMC (wt%)	Chemical Class
Witco Corp.			
Witcolate D51-51	A	0.034	alkyl phenoxyether sulfate
Witcodet 100	Mixture, A/N	N/A	alkylbenzene sulfonate/ alcohol ethoxylate
Emcol CNP-110	A	N/A	polyoxy carboxylate
Witconol APS	N	N/A	alcohol alkoxylate
Witconate AOS	A	0.072	sulfonate
Emphos CS-1361	A	1.0*	polyoxy alkylated phosphate ester
Witconol 2722	N	N/A	alcohol ethoxylate
Witconol TD-100	N	N/A	alcohol alkoxylate
Witconol SN-70	N	0.5*	alcohol ethoxylate
Witconol 1206	N	0.5*	polyalkylene glycol ether
Emcol CBA-60	A	N/A	polyoxy carboxylate
Stepan Co.			
Makon-10	N	N/A	alcohol alkoxylate
Stepanate SxS	A	N/A	sulfonate
Vista Chemical Co.			
Vista C550-LAS (V2157E)	A	1.5×10^{-2}	alkylbenzene sulfonate
Alfonic 1412-60 (V2158E)	N	3.4×10^{-4}	alcohol ethoxylate
Novel II 1412-70 (V2160E)	N	1.5×10^{-3}	alcohol ethoxylate
Alfonic 1012-60 (V2159E)	N	9.7×10^{-4}	alcohol ethoxylate
Rhône Poulenc			
Alcodet MC2000	N	N/A	thioethoxylate
Olin Chemicals			
PolyTergent S-405LF	N	N/A	alcohol alkoxylate (linear)
PolyTergent S-505LF	N	N/A	alcohol alkoxylate (linear)
PolyTergent SL-42	N	N/A	alcohol alkoxylate (linear aliphatic)
Union Carbide			
Terigtol			
- 15-S-9	N	0.0056	polyalkylene glycol ether
- 15-S-7	N	0.0039	alcohol ethoxylate
- XL-80N	N	0.0086	alcohol alkoxylate
- TMN-10	N	0.094	alcohol ethoxylate
- TMN-6	N	0.058	alcohol ethoxylate
- MinFoam 1x	N	0.0035	polyalkylene glycol ether
- MinFoam 2x	N	0.0019	polyalkylene glycol ether

Notes: * - CMC estimated from surface tension measurements.
CMC - Critical micelle concentration.
N/A - Not available.

which adhered to the vial sidewalls, causing variable and elevated turbidity readings unrelated to aqueous dispersion of hydrocarbons. However, results did not indicate soil colloid formation was likely, nor were adverse reactions between site soil or water constituents (such as precipitation or inactivation) observed. Undesirable effects such as sulfide formation in test vials and precipitation of surfactant solutions upon storage were noted during screening tests, and served as the basis for disqualification of 12 surfactants from Phase II testing.

Phase II screening tests were performed as batch washing studies. The first study was designed to quantitatively evaluate detergency of the remaining 16 surfactants and unamended tap water. This test was conducted by dispensing soil and surfactant solutions (0.1 percent, v/v) into beakers which were mixed on a gyrotary shaker for 30 minutes at 150 rpm, removed and allowed to settle, decanted, and rinsed with fresh surfactant and tap water. Treated soil was then analyzed for residual TPH.

TPH reduction ranged from 40 to 78 percent. Results indicated four surfactants, Alcodet MC2000 (MC2000), Emcol CBA-60 (CBA-60), Witcodet 100, and Witcolate D51-51 (D51-51), efficiently removed soil contamination. A single sample was re-rinsed with tap water and TPH reanalyzed. This result suggested the rinse procedure was inadequate to remove residual surfactant.

A second batch-washing test was designed to optimize the rinse process and surfactant concentration. This test was run using 0.5 and 1.0 percent v/v solutions, increasing the volume and number of rinses, and retesting the four most promising surfactants identified earlier. Results (Table III) indicate TPH reductions of 77 to 90 percent, with no substantial difference between the two tested concentrations, and support the conclusion that the original rinse procedure was inadequate to remove residual surfactant.

In both batch-washing tests, most soil fines were lost during the rinse procedure. This was thought to be due to the application of mechanical energy and surfactant activity, releasing fines previously adsorbed onto larger particle surfaces. The effect of loss of fines on TPH concentrations in the remaining soil may be significant.

Based on Phase I and Phase II results, these four surfactants were considered to have the most desirable characteristics and were selected for further testing in soil column studies.

Soil Column Testing. Soil column leaching studies were performed to further evaluate effectiveness of tap water and the four surfactant solutions previously identified. Study controls included soil columns flushed with tap water alone and soil columns which were assembled and sealed but not flushed. Flushing solutions were percolated through duplicate columns and leachate samples collected. Columns were then rinsed with tap water to remove surfactant residue and rinsate samples were collected from each treated column. After rinsing, a treated soil leachate sample was collected, and columns were then decommissioned. Treated soil was extruded from the columns and soil samples collected.

Table III. Laboratory Screening - Second Batch-Washing Test Results

Surfactant	Concentration (%, v/v)	Soil TPH (mg/kg)	% TPH Reduction
Emcol CBA-60			
2 x 50 ml rinse	0.5	4,800	79
2 x 100 ml rinse	0.5	3,000	87
4 x 100 ml rinse	0.5	5,400	77
2 x 50 ml rinse	1.0	3,300	86
Alcodet MC2000			
2 x 50 ml rinse	0.5	3,400	85
2 x 100 ml rinse	0.5	2,200	90
4 x 100 ml rinse	0.5	2,400	90
2 x 50 ml rinse	1.0	2,700	88
Witcodet 100			
2 x 50 ml rinse	0.5	2,600	88
2 x 100 ml rinse	0.5	2,600	88
4 x 100 ml rinse	0.5	2,600	88
2 x 50 ml rinse	1.0	2,700	88
Witcolate D51-51			
2 x 50 ml rinse	0.5	4,500	80
2 x 100 ml rinse	0.5	4,500	80
2 x 50 ml rinse	1.0	3,300	86
4 x 100 ml rinse	1.0	4,600	80
Tap Water			
2 x 50 ml rinse	NA	8,600	63
2 x 50 ml rinse[a]	NA	5,000	78

Notes: a - Test was rerun with a fresh soil sample.

Leachates were screened for TPH, VOCs, and PAHs at various times during the study; rinsates were screened for TPH. Treated soil was analyzed for TPH, VOCs, PAHs, and moisture content after leaching and rinsing procedures were completed. TPH/IR results were used as an overall indicator of contaminant reduction during testing.

Procedure. A set of 12 glass columns measuring 18 by 2-1/2 inches was packed with approximately 1,500 grams (g) of composite contaminated soil. Surfactant solutions were tested at 0.5 percent v/v concentration. Columns were flushed at the same rate, approximately 1 pore volume (delivered by a Masterflex peristaltic pump equipped with No. 16 pump heads) over a 4-hour period for 14 consecutive days. This flushing rate was selected to maximize contact time between target contaminants and solutions without saturating the soil. Column leachates were collected separately in sealed systems and stored at 4°C until surfactant flushing was completed; composited samples were then collected. Flushing solutions were not recycled. Ambient temperature, pH, and conductivity of column leachates were measured daily.

Treatment Leachate Sample Collection. Discrete samples for TPH analysis were collected after approximately 1, 7, and 14 pore volumes of flushing solution had been delivered and collected. Composited leachate aliquots were analyzed for VOCs, TPH, and PAHs.

Rinsate Sample Collection. After flushing, soil columns were rinsed with tap water, including the columns flushed only with unamended tap water. Tap water flushed columns were rinsed in order to keep total volumes flushed through each column comparable. The rinse was delivered at the same rate as flushing solutions, and rinsate was collected in the same manner. Columns were rinsed until residual surfactant had been removed. Surfactant presence in the rinsate was evaluated by a simple foam test.

Rinsing requirements ranged from 9 days (MC2000) to 40 days (Witcodet 100), at which time rinsing was arbitrarily discontinued due to time constraints. Rinsate collected from each column was composited and aliquots removed for TPH analysis.

Treated Soil Leachate Sample Collection. After rinsing, treated soil leachate samples were collected. These samples were expected to be indicative of water quality once soil had been flushed. Tap water was flushed through the columns and collected as before, using the minimum amount of flushing to satisfy volume requirements for analysis (approximately 10 pore volumes). All composite samples were analyzed for VOCs, TPH, and PAHs.

Treated Soil Sample Collection. Once sufficient treated soil leachate had been collected, columns were decommissioned and soil removed. Flushed soil was composited and samples collected for analysis of VOCs, TPH, PAHs, and moisture content. The two control columns which were not flushed were similarly decommissioned at this time and analyzed for the same constituents.

Results - Daily Monitoring. Ambient temperature, leachate pH and conductivity, and flushing solution flow rate were monitored daily. Ambient temperatures during the study ranged from 16.4 to 25.0 °C. This temperature range did not appear to have any impact on treatment. The pH of all column leachates ranged between 7 and 8, which approximately encompassed native soil and water pH. The mean total leachate collection volume was 4,940 +/- 380 ml. Leachate conductivity declined steadily in all columns as flushing continued. In general, columns flushed with Witcodet 100 had the highest conductivity. Initially, all column leachates were turbid, due in part to the chemical nature of the surfactant and surfactant/hydrocarbon micelles, but also due to mobilized fines. Fines were visible in all leachates and appeared to be washed out of the soil by the end of the surfactant flushing phase of this study.

Sulfide formation, indicative of the presence of sulfate which has been reduced under conditions of low redox potential, was evident in several soil columns. Darkening of soils, characteristic of metal sulfide formation, was most noticeable in the Witcodet 100 columns, but was also seen in the MC2000 columns. The D51-51 composite leachate showed evidence of sulfide production. The MC2000 leachate also had a visible nonaqueous-phase liquid layer during the first 5 to 6 days of flushing, and in the composite collection bottle.

Treatment Leachate - TPH Results. Results of TPH monitoring (Table IV) indicate the kinetics and amount of TPH removal varied between treatments. The highest TPH concentration was seen in the Witcodet 100 sample collected after the first pore volume had been flushed. All tap water leachates contained low TPH concentrations.

Table IV. Soil Column Testing - TPH Monitoring Results

Flushing Solution	TPH (mg/L) 1st Pore Volume	7th Pore Volume	14th Pore Volume
Emcol CBA-60	88	210	125
Alcodet MC2000	340	180	24
Witcodet 100	2,000	280	21
Witcolate D51-51	38	84	74
Tap Water	13	6	4

Notes: mg/L - milligrams per liter

Treatment Leachate - Composite Sample Results. Chemical analysis results are shown in Table V. TPH concentrations in composite samples were consistent with those measured in discrete pore volume samples except for Witcodet 100. This composite TPH value appears to be biased low, possibly an artifact caused by formation of stable hydrocarbon-surfactant micelles which were not effectively extracted or measured.

The highest concentrations of organic compounds were generally seen in the Witcodet 100 and MC2000 samples. All VOCs (except ethylbenzene) and PAHs detected in soil during the baseline characterization were detected in at least one treatment leachate. Tap water leachates contained low concentrations of all analytes; concentrations of organic constituents in surfactant treatments were 1 to 4 orders of magnitude higher than in tap water leachates.

Rinsate TPH Results. The volume of tap water required to rinse soils free of surfactant varied from approximately 3,600 to 17,900 ml. Rinsing of the Witcodet 100 columns was discontinued after 40 pore volumes. At this time, foam testing indicated surfactant residue was still present, although significantly reduced.

Table V. Soil Column Testing - Treatment Leachate Results

Analyte (ug/L)	CBA-60	MC2000	Witcodet 100	D51-51	Tap Water
TPH (mg/L)	29.9	186	30.2	7.1	1.4
VOCs					
1,2-Dichlorobenzene	1,200	1,950	2,200	860	74
1,4-Dichlorobenzene	120	330	300	100	13
cis-1,2-Dichloroethene	ND	ND	ND	ND	24
Toluene	ND	ND	140	ND	ND
Total Xylenes	210	580	510	ND	30
PAHs					
2 to 3 ring	166.2	610.4	508.8	94	0
4 to 6 ring	496.5	1,916.2	862.2	463.8	0.16

Notes: ND - Not detected.　　　ug/L - micrograms per liter.

Rinsate TPH (Table VI) was the lowest in tap water and CBA-60 rinsates and highest in Witcodet 100 rinsates. Witcodet 100 rinsates contained the highest TPH concentration despite their significantly greater total volume.

Table VI. Soil Column Testing - Rinsate TPH

Flushing Solution	TPH (mg/L)
Emcol CBA-60	1.1
Alcodet MC2000	3.0
Witcodet 100	9.7
Witcolate D51-51	5.8
Tap Water	1.5

Treated Soil Leachate Results. Chemical analyses (Table VII) were performed on treated soil leachates to evaluate the effects of soil treatment on leachate chemical composition. TPH concentrations were low to not detectable in all treatments. The D51-51 samples contained the same TPH concentration as rinsates, indicating these columns were likely not effectively rinsed. The same organic compounds detected in surfactant-flushed leachates were present. The Witcodet 100 samples generally contained the fewest analytes and the lowest concentrations, and tap water leachates contained the most analytes and highest concentrations.

Treated Soil Results. Table VIII summarizes treated soil results. Moisture content in flushed columns was comparable to and only slightly higher than the nonflushed control. TPH concentrations were similar in columns flushed with D51-51, CBA-60, tap water, and nonflushed controls. Flushing with these solutions did not appear to measurably reduce TPH concentrations when compared to the nonflushed control. TPH concentrations in the Witcodet 100 and MC2000 treatments indicated 92 and 82 percent reduction.

Table VII. Soil Column Testing - Treated Soil Leachate Results

Analyte (ug/L)	CBA-60	MC2000	Witcodet 100	D51-51	Tap Water
TPH (mg/L)	1.2	ND	0.8	5.6	0.7
VOCs					
1,2-Dichlorobenzene	115	53	4.8	70	135
1,4-Dichlorobenzene	13	7.0	1.5	14	11
cis-1,2-Dichloroethene	6.1	9.1	3.8	4.4	10
Toluene	ND	ND	1.1	ND	ND
Total Xylenes	34	14.6	ND	ND	34
PAHs					
2 to 3 ring	2.01	2.31	0.06	0.69	0.84
4 to 6 ring	1.62	2.46	0.14	0.91	2.00

Table VIII. Soil Column Testing - Treated Soil Results

Analyte (mg/kg)	CBA-60	MC2000	Witcodet 100	D51-51	Tap Water	Control[a]
Moisture Content (%)	7.45	7.73	7.38	8.51	8.24	5.15
TPH	13,000	2,900	1,400	17,000	15,000	16,000
VOCs						
1,2-Dichlorobenzene	3.3	0.66	0.18	3.8	9.5	7.8
1,3-Dichlorobenzene	ND	ND	ND	ND	0.15	ND
1,4-Dichlorobenzene	0.8	0.098	0.031	0.88	1.1	1.2
cis-1,2-Dichloroethene	0.20	0.12	0.74	0.36	0.12	ND
Ethylbenzene	ND	0.18	0.28	ND	0.81	ND
Toluene	0.34	0.11	0.22	ND	0.19	ND
Total Xylenes	1.0	0.48	0.58	0.56	1.8	1.6
PAHs						
2 to 3 ring	3.97	0.98	0.21	4.53	6.56	9.34
4 to 6 ring	32.85	7.71	2.27	38.01	47.34	56.95

Notes: a - Control columns were not flushed.

Generally, the highest concentrations of organic compounds were seen in either the tap water flushed or nonflushed samples, and the lowest in the MC2000 and Witcodet 100 samples. VOC concentration reductions ranged from 0 to 98 percent for these two treatments, and PAH reductions ranged from 49 to 99 percent for individual compounds.

Summary of Results - Soil Column Testing. Flushing soil with tap water proved to be minimally effective at mobilizing TPH, VOCs, and PAHs, based on comparison to nonflushed control samples. Surfactant solutions demonstrated superior performance to tap water. The most effective surfactant for mobilization of TPH, VOCs, and PAHs was Witcodet 100 (an anionic/ nonionic blend from Witco Corp.), with MC2000 (an anionic product from Rhône Poulenc) only slightly less effective. TPH reductions up to 92 percent and VOC and PAH reductions up to 99 percent were observed.

Although MC2000 and Witcodet 100 appeared to be approximately equally effective at contaminant mobilization, their behavior in the test systems was markedly different. The Witcodet 100 leachates were stable and emulsified TPH remained in solution. Alcodet MC2000 leachates were unstable during the first half of the test, and became biphasic within several hours of collection.

Additional Leachate Characterization. Additional studies focused on regulated parameters of concern to the local Publicly Owned Treatment Works (POTW) and Water Pollution Control Department, and activity of an emulsion-breaking product on MC2000 and Witcodet 100 leachates. Results were consistent with soil column testing data. The concentration of emulsion-breaking additive required was significantly greater than ordinarily used, and MC2000 leachates were only treatable when diluted by a factor of 7 or more.

Summary and Discussion

Twenty-eight surfactants were screened to evaluate their relative ability to emulsify motor oil and hydrocarbons adsorbed onto soil. Undesirable effects such as sulfide formation and surfactant precipitation were noted. Four surfactants identified during the screening process were evaluated along with tap water in a soil column flushing study.

Column study data indicated two surfactants, Alcodet MC2000 and Witcodet 100, were highly effective in removing petroleum hydrocarbons and other organic compounds from soil. These surfactants facilitated removal of 80 to 90 percent of the TPH and 49 to 99 percent of the PAHs relative to untreated control samples after approximately 14 pore volumes of flushing. Tap water treatment resulted in reduction of TPH by approximately 6 percent and PAHs by 0 to 37 percent. Tap water treatment did not reduce the concentration of any 4-, 5-, or 6-ring PAHs, with the exception of indeno(1,2,3-c,d)pyrene.

Several observations and considerations offset the success of surfactants in reducing overall contaminant load. Issues important to the evaluation of surfactant-based soil flushing for this site are as follows:

- Mobilization of soil fines: Mobilization of soil fines along with hydrocarbons was observed during soil flushing studies. This observation raises a general concern over potential occlusion in the subsurface due to mobilized fines, stable surfactant/contaminant micelles, or hydrolyzed surfactant flocs during full-scale soil flushing.

- Surfactant recovery: Study data indicate surfactant rinsing from soil was variable; 40 pore volumes of rinse water was insufficient to completely remove Witcodet 100. The fate and effects of residual surfactant are unknown, but it is expected that low-level leaching could occur as long as surfactant residue was present. These considerations have significant impact on full-scale implementation at the site, including regulatory approval and long-term protection of groundwater quality.

- Surfactant stability: A critical observation made during the study reflected the potential for surfactant/contaminant micelles of Alcodet MC2000 to disaggregate, forming two phases. This observation raises the general question of surfactant/contaminant stability and subsequent recovery from the subsurface, which may be affected by dilution of the flushing solution with groundwater prior to recovery and treatment.

- Microbial transformation of surfactants: Alteration of subsurface redox conditions (i.e., creation of anaerobic conditions) is an important consideration in evaluating full-scale implementation. Limited data are available concerning the biological degradation of surfactants. Two concerns are the formation of heavy metal sulfides from surfactants containing sulfonate functional groups and creation of general anaerobic conditions through oxygen consumption during surfactant degradation.

- Leachate recovery: Leachates containing high concentrations of contaminants would be created during the early stages of soil flushing. Groundwater movement at the site, as assessed by hydraulic conductivity measurements, is extremely rapid. If a failure in the groundwater recovery system were to occur, a slug of highly contaminated water would be released.

- Treatability of soil flushing leachate: During full-scale implementation, soil flushing leachate would be captured, extracted from the subsurface, and pretreated on site prior to discharge to the city POTW. The initial pretreatment step would be effluent de-emulsification to facilitate oil/water separation by dissolved air flotation techniques. Evaluation of surfactant-generated leachates indicates potential technical difficulty and significant cost associated with reversing their emulsifying effects.

- Cost: The quantity of surfactant required for full-scale implementation would be considerable since the soil would require flushing, with a minimum of several pore volumes of surfactant solution.

In summary, surfactant-based flushing solutions were shown to effect significant reduction of organic contaminants present in site soils. However, counter-indications for surfactant use were observed. Additionally, available fate and toxicity data are insufficient for the majority of products tested. The use of surfactant-based flushing solutions was therefore not recommended for this site.

LITERATURE CITED

1). Nash, J.; Travers, R.P. Field Evaluation of In Situ Washing of Contaminated Soils with Water/Surfactants, *Proceedings of the Twelfth Annual Research Symposium on Land Disposal, Remedial Action, Incineration, and Treatment of Hazardous Waste*; EPA/600/9-86/022; United States Environmental Protection Agency: Cincinnati, OH, 1986, pp 208 -217.

2). Vignon, B. W.; Rubin, A. J. Practical Considerations in the Surfactant-aided Mobilization of Contaminants in Aquifers, *JWPCF, 1989, vol. 61, No. 7*, pp 1233 - 1240.

3). Rickabaugh, J.; Clemment, S.; Martin, J.; Sunderhaus, M.; Lewis, R. F. Chemical and Microbial Stabilization Techniques for Remedial Action Sites, *Proceedings of the Twelfth annual Research Symposium on Land Disposal, Remedial Action, Incineration, and Treatment of Hazardous Waste*; EPA/600/9-86/022; United States Environmental Protection Agency: Cincinnati, OH, 1986, pp 193 - 200.

4). Abdul, A. S.; Gibson, T. L.; Rai, D. N. Selection of Surfactants for the Removal of Petroleum Products from Shallow Sandy Aquifers, *GW, November - December 1990, Vol. 28, No. 6*, pp 920 - 926.

5). Abdul, A. S.; Gibson, T. L.; Ang, C. C.; Smith, J. C.; Sobczynski, R. E. In Situ Surfactant Washing of Polychlorinated Biphenyls and Oils from a Contaminated Site, *GW, March - April 1992, vol. 30, No. 2*, pp 219 - 231.

6). Shiau, Bor-Jier; Sabatini, D. A.; Harwell, J. H. Solubilization and Microemulsification of DNAPLs Using Edible Surfactants, submitted to *GW*, 1993.

7). U. S. Environmental Protection Agency, *Test Methods for Evaluating Solid Waste, Physical\Chemical Methods*; SW-846; Office of Solid Waste and Emergency Response: Washington D. C., 1986, 3rd Edition.

8). U. S. Environmental Protection Agency, *Methods for Chemical Analysis of Water and Waste*; EPA 600/4-79-020; U. S. Environmental Monitoring and Support Laboratory: Cincinnati, OH, March 1983.

9). American Petroleum Institute, Underground Movement of Gasoline on Groundwater and Enhanced Recovery by Surfactants, *API Publication No. 4317*, Health and Environmental Sciences Department: Washington, DC., September 1979.

RECEIVED December 13, 1994

Chapter 13

Enhanced Removal of Dense Nonaqueous-Phase Liquids Using Surfactants
Capabilities and Limitations from Field Trials

John C. Fountain, Carol Waddell-Sheets, Alison Lagowski, Craig Taylor, Dave Frazier, and Michael Byrne

Department of Geology, State University of New York, 772 Natural Sciences and Mathematics Complex, Buffalo, NY 14260

Results of two pilot field tests suggest that surfactant flushing can be successful under the following conditions: 1) free phase or residual DNAPL is present; 2) the hydraulic conductivity is moderate to high (> 10^{-3} cm/sec) and 3) an aquitard is present below the target zone to act as a barrier to vertical migration. These tests indicate that hydrogeologic parameters (aquifer heterogeneities) and contaminant distribution not surfactant performance are the variables that determine the ultimate level of remediation of DNAPL sites. Surfactant performance parameters affect the time and cost required to reach a specific remediation level. Surfactant performance depends upon surfactant type and concentration, contaminants present, water chemistry and aquifer materials (clay content and types, and organic content). Field parameters include the geometry of injection and extraction systems, contaminant distribution and aquifer heterogeneities. These results provide guidance for surfactant selection and for the determination of realistic remediation goals using surfactant enhanced technology.

The principal limitations of standard pump and treat remediation is their inability to mobilize most hydrophobic organic solvents or dense non-aqueous phase liquids (DNAPLs) except as dissolved phase. Because DNAPLs have relatively low aqueous solubilities (hundreds to several thousand ppm) and significant sorption within many sedimentary systems (1-9), these organic compounds are relatively immobile in the subsurface and hence are not conducive to rapid extraction by aqueous dissolution. In addition, as they slowly dissolve, DNAPLs act as continuous, long-lived contamination sources which will persist for decades if not centuries and thus will require extensive time frames for remediation unless these DNAPL sources can be removed (8,10-15).

0097–6156/95/0594–0177$12.00/0
© 1995 American Chemical Society

Surfactants can address this problem by both aiding in locating and remediating DNAPL source zones. Surfactants increase the mobility of the contaminants by combinations of the following three mechanisms: 1) increasing the solubility of the contaminants (4,16-21); 2) reducing their sorption (22,23) and 3) lowering high interfacial tensions (IFT) between water and DNAPLs (19,24). After removal of the DNAPL source, the plume can then be remediated by standard pump and treat.

Surfactant enhanced remediation is cost effective for DNAPL sites because the DNAPL zones are generally restricted in size. Due to their high densities, DNAPL motion is primarily controlled by the interaction between gravity and capillary forces, not by horizontal groundwater flow (11,25-28). Thus DNAPLs move downward until a low permeability zone is encountered. This results in minimal horizontal spreading and produces pools, lenses and isolated ganglia of DNAPL (10,29). The dissolved plume however, may extend thousands of feet from the original site of the spill or leak as it is transported by the groundwater flow.

Field Demonstration of Surfactant Enhanced DNAPL Remediation

To date, two successful field trials have been conducted using surfactant enhanced pump and treat remediation of DNAPL contaminated aquifers under radically different aquifer conditions. The first trial was conducted from June 1990 to August 1991 at the Canadian Forces Base Borden (CFB) in a clean sand with <1% clay, <1000 mg/kg organic content and fresh groundwater. This test occurred in a 3 m x 3 m x 3 m cell which had been contaminated with 271 liters of PCE in a controlled release. The second trial was conducted from June 1991 to February 1993 at a chlorocarbons manufacturing plant in Corpus Christi, Texas in a fine-grained sand with variable smectitic clay content (1-15%), little organic carbon (250 - 310 mg/kg) and highly saline groundwater (12,000 ppm total dissolved solids).

These sites share the following characteristics: 1) the presence of a well delineated DNAPL zone; 2) a small area of concern ($\leq$10 m x 10 m x 10 m); 3) easy access to the test area; 4) moderate hydraulic conductivity ($\geq 10^{-3}$ cm/sec) and 5) the presence of a thick clay aquitard beneath the target zone to maintain hydraulic control.

Core and groundwater analyses from these tests suggest that when surfactants are used, DNAPL mass removal progresses at a rate considerably faster than what would be expected in standard pump and treat techniques. Core data in Table I suggests the following: 1) the surfactant solution rapidly reduced the amount of DNAPL present especially in the zones of higher hydraulic conductivity; 2) contaminated zones showed marked reductions in both maximum concentration and thickness of DNAPL; 3) pools of DNAPL remained at the same elevation throughout both tests indicating little vertical migration occurred due to lowered interfacial tensions and 4) remediation was incomplete at both sites because DNAPL remained in zones of low hydraulic conductivity (Table I). The upper PCE zone at Borden (0-1 m BGS) was perched on a layer within the sand less than 2 cm in thickness. This zone was

Table I. Core Data for Field Trials at CFB Borden and Corpus Christi, Texas					
Parameter	Depth BGS[b]	Initial Value	Final Value	Pore Volumes	Comments
CFB Borden					
PCE Residual saturation[a]	0-1 m	10%	<1%	14.4	SAND no visible difference at 2 cm scale
PCE Pool Height	2.5-3 m	50 cm	2-3 cm	14.4	Perched on clay aquitard
PCE Pool saturation		20%	3%	14.4	
Corpus Christi, Texas					
CTET Concentration[c]	10-12 ft	>2000[d]	>2000	3	Clay
			>2000	12.5	
CTET Concentration	12-14 ft	77-2956	>2000	3	Clayey Sand 10-30% cl
			<10	12.5	
CTET Concentration	16-24 ft	574-2674	<10	3	Sand with 1-5% clay
			<10	12.5	

[a] Saturation is the ratio of the volume of PCE in the sample to the pore volume of the sample.
[b] Below ground surface
[c] All concentrations are in mg/L
[d] Maximum concentration

similar in hydraulic conductivity and appearance to the rest of the section. The lower pool was perched on the clay aquitard at the base of the cell. The CTET at Corpus Christi was located within three horizons, a clay zone from 10-12 feet BGS, a clayey (10-30%) sand from 12-14 feet BGS and a sand zone from 16-24 feet BGS which was underlain by a clay aquitard at 24 feet BGS.

Monitoring and extraction well data also indicate rapid remediation of DNAPL zones had occurred. At Corpus Christi, effluent carbon tetrachloride (CTET) concentrations dropped from greater than 1000 ppm to less than 10 ppm after three pore volumes of surfactant were circulated. After 12.5 pore volumes of surfactant were circulated, a total of approximately 276 L of CTET was removed. At Borden, effluent PCE concentrations dropped from 4000 ppm to 200 ppm in 10 pore volumes and remained at approximately 200 ppm at the time the test was terminated. After 14.4 pore volumes of surfactant were circulated, 62 liters of tetrachloroethylene (PCE) were extracted whereas only 12 liters were extracted after 6.2 pore volumes of standard water flushing.

The results of these field trials suggest that use of surfactants is an effective method for rapid reduction of DNAPL zones but hydrogeologic parameters determine the remediation levels achievable at a given site. Hydrogeologic considerations include both the limitations inherent in standard pump and treat systems (delivery considerations) and limitations specific to surfactant enhanced systems.

General Limitations of Pump and Treat Systems

Because using surfactants to enhance contaminant solubility is essentially a modified pump and treat approach, surfactant remediation shares the limitations inherent in such systems (3,5,30). These limitations include any geologic condition which will limit the flow of groundwater through the target zone. The principal limitation is of course, the hydraulic conductivity (K) of the target zone. Hydraulic conductivities on the order of 10^{-3} cm/sec are necessary for sufficient pumping rates to make using pump and treat a viable method (5). While subsurface materials with hydraulic conductivities lower than this can be remediated, the time and costs required become prohibitive.

Other parameters within the system will limit the remediation of specific zones within the target interval and cause incomplete remediation or tailing. Hydrodynamicly isolated "dead spots" will result in areas that cannot be remediated easily since the only transport mechanism available is diffusion which is extremely slow (3,9,30). For example, dead end fractures in a rock matrix or fractured clays cannot be treated readily using pump and treat methods since DNAPL cannot be physically pumped upward once it has entered a fracture. Impermeable or low permeability lenses will create shadow zones down gradient that will also clean slowly, if at all since groundwater flows around these areas. In addition, DNAPL that has penetrated such lower permeability zones will diffuse slowly into adjacent higher permeability layers, acting as a continuous long term source of contamination (3,5,30) and the groundwater flow through these low K zones will also be slower retarding the remediation of these zones.

Another factor that controls the rate of remediation in unconsolidated aquifers is fluctuations in groundwater table elevations. Lowered water tables may restrict the hydraulic gradient that can be induced if the saturated thickness of the sediments has been reduced significantly. This will result in lower pumping rates than expected and may require a closer-spaced injection/extraction array. In addition, a lower water table will isolate pockets of DNAPL above the pumping zone, leaving DNAPL to contaminate the remediated zone when the water table returns to normal. This effect is similar to the presence of residual DNAPL in the vadose zone which acts as a source of new contamination each time rainwater percolates down through a contaminated zone.

The presence of any of these characteristics will result in slower and possibly incomplete remediation of a target zone when a pump and treat system is employed. The presence of surfactant in the system may speed the remediation of the low permeability zones to some extent by lowering the interfacial tension and allowing the penetration of these zones by surfactant laden groundwater but these zones will still slow the process as a whole.

Limitations of Surfactant Enhanced Systems

In addition to the general limitations inherent in any pump and treat system, a surfactant enhanced system has several other limitations. These limitations include solubilization limits, sorption, surfactant degradation, water chemistry, DNAPL mobilization, and site characterization.

Surfactant Solubilization Limits. The increase in solubility provided by surfactants for any given contaminant is a function of the aqueous solubility of the compound (i.e. the higher the aqueous solubility, the lower the attainable solubilization increase). For example, the solubility of 1, 2-dichloroethane with an aqueous solubility of 8690 ppm increases by only a factor of three in a 1% surfactant solution; whereas, the solubility of PCE (150 ppm) increases by a factor of 135 in a 1% surfactant solution. While this effect limits the increase in efficiency that can be obtained directly by solubilization for more soluble organic compounds, other effects such as enhanced desorption of contaminants combine to produce significant increases in extraction efficiency (*22,23*).

The presence of multi-component DNAPLs also imposes limitations on surfactant performance. Generally, compounds with significantly different polarities as reflected in their aqueous solubilities are not solubilized equally by any given surfactant. For example, when the solubilization of a mixture of 1% Aroclor 1254 and transformer oil is compared in a surfactant optimized for transformer oil and one optimized for PCBs different ratios between the mg/L transformer oil solubilized and the mg/L PCBs solubilized result. When using a surfactant optimized for transformer oil a ratio of 311 results but a ratio of only 15 results when the surfactant is optimized for PCBs (Byrne, M., SUNY at Buffalo, unpublished Master's thesis, 1993). Thus if DNAPL components have a wide range of aqueous solubilities, optimal solubility enhancement may not be achieved for each compound. In general, the most significant compound is targeted because minor components will still be removed.

Surfactant Sorption. Since surfactants are organic compounds sorption may be significant (*1,4,20,31,32*). This will reduce the efficiency of the system and increase the final cost. Surfactant sorption is a function of the surfactant type, total organic carbon content, water chemistry and the type and percentage of clays present in the soil (*4,20,31,32*, Lagowski, A., SUNY at Buffalo, unpublished Master's thesis, 1994). Generally, as organic content or smectitic clay content increases, sorption of surfactants also increases; however, the actual sorption for any given natural system must be determined because the amount of sorption is highly dependent upon surfactant type. In general, anionic surfactants have lower sorption than nonionic surfactants but also tend to have lower solubilization capacities (*20,33*, Lagowski, A., SUNY at Buffalo, unpublished Master's thesis, 1994). If sorption of the best solubilizer is high enough, however, a lower solubilizer with lower sorption may be more economical. The chemistry of the aquifer and its effects on surfactant sorption must be determined for each site using site specific soil and groundwater or soil and water chemistries that mimic the site conditions as closely as possible before final cost estimates can be made. Once laboratory tests have been completed, they should ideally be followed by pilot scale field tests to determine large scale effects.

Water Chemistry. The effects of water chemistry on the solubility of surfactants has been well established (*20,34*). Most surfactants with hydrophilic/lipophilic balances (HLB) above approximately 12-13 will readily dissolve in distilled water, however; ionic strengths of as low as several hundred ppm may cause the surfactant to precipitate, making it unusable. Thus site specific water chemistry must be used during testing to determine whether a given surfactant is suitable for use under site specific conditions.

Water chemistry also effects surfactant sorption. The composition of the groundwater affects the surface properties of the clays present such as cation exchange capacity and surface area especially when smectitic clays are present. For example, sodium smectite when altered to calcium smectite will show a decrease in surface area, cation exchange capacity and swelling capacity (*35*). Because surface area is a primary factor controlling the sorption of the surfactants, any change may result in erroneous estimations of site sorption parameters. Use of distilled water will mobilize any clays present in a system resulting in different sorption characteristics (Lagowski, unpublished Master's Thesis, SUNY Buffalo, 1994). Thus it is necessary to use water chemistry similar to the site to insure that clay properties have not been altered.

Surfactant Degradation. The biotic and abiotic (hydrolysis) degradation of surfactants is important because it destroys the surfactant, inhibiting its efficiency. In addition, biodegradation may cause biofouling of tanks and injection wells if uncontrolled. Laboratory experiments indicate that the efficiency of the surfactants may decrease by as much as 25% six months after mixing even when no visible evidence of degradation is observed. Aerobic and anaerobic experiments suggest that common surfactant classes also degrade rapidly within natural environments (*36-38*; Stewart, B., SUNY at Buffalo, unpublished Master's thesis, 1994). Modifications of the system design can be used to minimize biodegradation. Field trials have indicated that minimizing the retention time in the static holding tanks can significantly reduce biofouling.

DNAPL Mobilization. The possibility of inducing unwanted vertical mobility by decreasing the interfacial tension (IFT) (*39,40*) is the most significant risk associated with surfactant enhanced remediation. The potential exists for making DNAPL removal considerably more difficult, if not impossible. If the capillary forces are reduced sufficiently to allow enough horizontal motion to permit extraction, vertical motion will also probably occur causing contamination of lower zones.

As previously stated, DNAPLs movement through the subsurface is ultimately governed by capillary forces (*11,25-27*). To enter each water filled pore, the DNAPL must overcome a capillary displacement pressure (*26,41,42*). This displacement pressure is greater in fine-grained material which act as barriers to downward migration until sufficient DNAPL head develops to overcome the displacement pressure. DNAPL will generally occur as residual saturation along its travel path and as one or more pools perched upon fine grained layers.

Surfactants reduce interfacial tensions between water and DNAPLs (*43,44*; Lagowski, A., SUNY at Buffalo, unpublished Master's thesis). The reductions may range up to four orders of magnitude depending upon the surfactants used. Since displacement pressure decreases directly with interfacial tension, large reductions of interfacial tension may allow DNAPL to penetrate fine grained layers that previously acted as barriers. Thus large reductions in the interfacial tension may allow DNAPL movement through an aquitard to contaminate previously remediated zones.

In addition, horizontal movement of DNAPL will also increase the chance of encountering vertical pathways such as fractures, sand lenses, root holes and other potential pathways that may exist within an aquitard. The reduction in surface tension makes penetration of these vertical pathways more likely, thus increasing the risk of vertical mobility. Since natural aquitards are seldomly free of such pathways and are generally heterogenous, some risk of vertical DNAPL mobilization is inevitable when surfactants are used.

The risk of inducing vertical migration can be minimized by using surfactants that do not lower the interfacial tensions by more than one order of magnitude. Core data from the two field tests suggests that under field conditions, no vertical mobility occurred. Although the IFTs were reduced from 45 dynes/cm to 3 to 5 dynes/cm, both the PCE at Borden CFB and the CTET at Corpus Christi remained suspended in the sandy matrix after the introduction of the surfactant.

Due to the potential risk inherent in large reductions in interfacial tensions, we do not recommend attempting to mobilize DNAPLs by minimizing IFTs. This involves reducing the interfacial tensions by several orders of magnitude (*24,25*). For DNAPLs, this approach produces an unacceptably high risk of vertical mobilization. For LNAPLs which have densities less than water, vertical mobilization is not a concern unless DNAPLs are also present.

Site Characterization. Remediation of any contaminated site requires extensive knowledge of the site history and geology in order to insure the contamination is located and removed as efficiently as possible. To remediate DNAPL site, the chemical composition, location and the distribution of the DNAPL must be ascertained. In addition, if pump and treat technology is utilized, the geology of the target zone must be examined to determine the potential tailing effects that will result from contaminated zones with low hydraulic conductivity which will determine the ultimate level of remediation achievable.

DNAPL Zone Characterization. Any remediation technology that attempts to address the problem of DNAPLs, by either extraction or containment, must be applied to the section of the site that contains the DNAPL source zone. In the case of surfactant enhanced pump and treat systems, the injection/extraction arrays must be installed in geometry that will result in the efficient sweeping of the entire DNAPL source zone. This requires an accurate determination of the DNAPL distribution prior to remediation. Determination of the presence of DNAPL is non-trivial, however, since no reliable method for locating DNAPL has yet been found (8,10,12). Standard analyses from monitoring wells can be deceptive since distances of only a few tens of feet from a DNAPL source may be sufficient to lower the concentrations far below aqueous solubilities suggesting a lack of DNAPL (10).

While surfactants share with other technologies the limitation that the DNAPL source zone must be defined, surfactants offer the unique capability to provide additional information for the characterization of the source zone. Once preliminary site characterization using cores, soil gas surveys and monitoring wells has identified the approximate geometry of the DNAPL zone, single and multi-well surfactant-extraction tests can be used to better define the location and composition of the DNAPL zone (45). Such tests, often involving existing extraction or monitoring wells, also allow examination of specific surfactant solutions for later use during remediation. Using an existing well array, the surfactant is injected into one well and extracted at another. The presence of DNAPL in the effluent indicates that DNAPL is present along the flow path between the two wells. In a single well injection test, the surfactant is injected and then extracted for a given period of time. If analysis of the effluent indicates higher concentrations of DNAPL than was obtained during water flushing, DNAPL is present. The length of time required for the DNAPL to appear is proportional to the distance from the well. The longer the time the farther away the DNAPL is. Of course the distance measured is determined by the length of the test, hydraulic conductivity of the sediments and the sorption capacity of the sediments. The number and spatial distribution of single well tests performed will determine the accuracy with which the DNAPL can be located.

The use of surfactants to delineate the DNAPL source zone also provides a powerful tool for DNAPL characterization. Due to the combination of dilution effects and the low solubilities of minor components in multi-component DNAPLs, minor components may often be present in water at concentrations below the limit of detection. Since surfactants increase the solubilities of low solubility compounds by up to several orders of magnitude, minor components may be identified by surfactant injection tests that would otherwise be missed. For example, PCB solubility may be increased by more than 10,000 times (Byrne, M., SUNY at Buffalo, unpublished data). Therefore, to determine whether PCBs are present, a surfactant that optimizes PCB solubilization can be injected initially. If the concentration of PCBs in the surfactant effluent increases significantly over the aqueous phase concentration PCBs are present as a component of a DNAPL phase. By repeating this test for several target compounds or classes of compounds, the DNAPLs composition can be more accurately determined than would be possible if only water is used. To design an efficient treatment system, knowledge of contaminant composition is critical.

Geologic Heterogeneities. Core analysis from the field trials have shown that surfactants can rapidly decrease DNAPL contamination in zones of high hydraulic conductivities (K); however, effluent concentrations may still remain at several tens to hundreds of parts per million after these zones are treated (Figure 1). This tailing results in part because contamination persists in lower K zones due to slower flow rates within these zones. This effect can be seen at both Borden and Corpus Christi. At Borden, 55 liters of PCE were removed within the first 10 pore volumes and only 7 were removed from 10 to 14 pore volumes. At Corpus Christi, monitoring wells NW1A and NW1B show different remediation rates apparently due to variations in hydraulic conductivities. The aquifer at Corpus Christi is a sand lense within a regional clay unit. It is composed of two sections: an upper zone of clayey sand, 4-7 feet thick with 10-30% smectite clay and a lower zone composed of a fine, well sorted sand with 1-5% smectite clay from 8-9 feet thick. Core analyses indicate that both zones contain DNAPL. Well NW1A monitored the upper zone (12-15 feet below ground surface) and NW1B the lower zone (21-24 feet BGS). The initial concentrations in the monitoring well were 256 and 210 ppm respectively at the start of the test (Figure 2). When surfactant reached 0.5% in the effluent (breakthrough), the CTET concentrations jumped to 834 and 860 ppm respectively. Within one month after surfactant flushing began, the monitoring well concentrations had decreased to 348 and 128 ppm respectively indicating the rapid removal of DNAPL. As expected from the higher clay contents and thus lower K, the upper unit cleaned more slowly.

The effect of different K zones leads to stepped concentrations in monitoring and extraction wells (Figures 2, 3). The initial phase of extraction would encompass the mass removal of DNAPL from the high permeability zones engendering a steep rise in the effluent concentrations (Figure 3). As remediation progresses and the surfactants penetrated the lower velocity sand zones, effluent concentrations would increase, reflecting the addition of DNAPL from this second zone. The concentrations would then decrease as the DNAPL from zone B was exhausted and would decrease again as the DNAPL

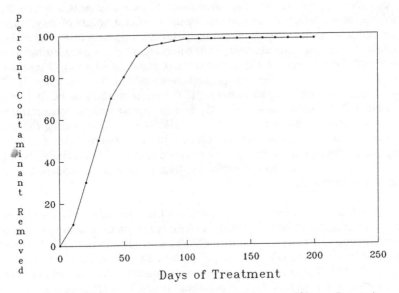

Figure 1. Generalized DNAPL extraction profile. Percent contaminant removed is plotted versus treatment time.

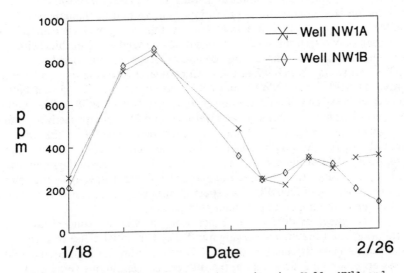

Figure 2. CTET Concentrations in Monitoring Wells NW1A and NW1B, Corpus Christi, Texas versus time.

in A was exhausted until a tail results which is caused by the contaminant trapped in clays and silts which have the lowest K. This type of effect is visible in the data from monitoring wells NW1A and NW1B at Corpus Christi. In both wells, CTET concentrations increased again after an initial steep declines in CTET concentration (Figure 2). These increases probably represent the arrival of DNAPL extracted from zones of lower hydraulic conductivity within each unit or dissolution of DNAPL pools still in place.

Surfactant Flood Design

Design of a field test requires, at a minimum, obtaining several cores to provide materials for aquifer characterization (fraction organic content and clay amount and types, spatial variability) and for column testing to derive mass transfer relationships and surfactant retardation factors. The aquifer material can also be used to measure the relationships between relative permeability and the capillary pressure and the residual NAPL saturation. With such relationships, and the data on surfactant performance, it is possible to design optimal surfactant floods and to predict the progress of surfactant-enhanced aquifer remediation by numerical methods (*14*, Brown, C.L.; Pope, G.A.; Abriola, L.M.; Sephrnoori, K. *Water Res, Res.* in press).

Conclusions

Surfactant enhanced pump and treat systems can remove a portion of the DNAPL mass rapidly and cost effectively but the ultimate remediation level is governed, as expected, by the hydrogeology of the site. The performance of the surfactant will affect the rate at which mass is removed from the aquifer and is controlled primarily by contaminant solubilization and surfactant sorption. The hydrogeology determines the extent of tailing and hence the ultimate remediation levels that can be achieved at a given site. Since aquifer heterogeneities are primarily responsible for tailing, a thorough understanding of the hydrogeology of the site is required to ascertain these limits. In addition to minimize costs, an accurate knowledge of the DNAPL distribution is necessary so that the entire source zone is treated while minimizing the treatment of non-DNAPL zones.

The results from both Borden and Corpus Christi indicate that surfactant flushing can be useful under a wide range of aquifer conditions as long as hydraulic conductivity is high enough (greater than 10-3 cm/sec) to make pump and treat a viable method.

Determination of the best surfactant for use is dependent upon the following criteria: 1) solubilization of the contaminant; 2) minimization of decreases in IFT to limit vertical mobility; 3) minimization of surfactant absorption and 4) surfactant toxicity and biodegradation. The potential for vertical migration is the primary risk associated with the use of surfactants and thus interfacial tensions should be an important criteria for surfactant selection.

Due to the numerous variables involved, quantitative prediction of the effect of aquifer heterogeneities on surfactant remediation requires numerical

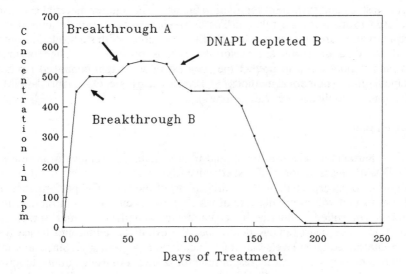

Figure 3. The effect of geologic heterogeneities on contaminant effluent concentrations.

modeling of each site. Due to the potential of DNAPL mobilization, a multiphase flow model should ideally be used which is coupled with the site characterization data obtained in earlier phases of the study.

Acknowledgments

The Corpus Christi research was sponsored by the New York Center for Hazardous Waste Management and Du Pont Corporation. The Borden Field trial was partially supported by the Solvents in Groundwater Program of the Waterloo Centre for Groundwater Research.

Literature Cited

(1) Karickhoff, S.W.; Brown, D.S.; Scott, T.A. *Water Res.* **1979**, *13*, 241-248.
(2) Gabarini, D.R.; Lion, L.W. *Envir. Sci. and Technol.* **1986**, *20*, 1263-1269.
(3) Keely, J.F. *Performance Evaluations of Pump-and-Treat Remediations* EPA/540/4-89/005: Office of Research and Development, U.S. EPA: Washington, DC, 1989.
(4) Liu,Z.; Laha, S.; Luthy, R.G. *Water Sci. Tech.* **1991**, *23*, 475.
(5) Hall, C.W.; Johnson, J.A. *J. Haz. Mat.* **1992**, *32*, 215-223.
(6) Imhoff, P.T.; Jaffe, P.R.; Pinder, G.F. *Water Res. Res.* **1993**, *30*, 307-320.
(7) Lesage, S.; Brown, S. *J. Cont. Hydr.* **1994**, *15*, 57-71.
(8) Whelan, M.P.; Voudrias, E.A.; Pearce, A. *J. Cont. Hydr.* **1994**, *15*, 223-237.
(9) Harmon, T.C.; Roberts, P.V. *Env. Prog.* **1994**, *13*, 1-8.
(10) Anderson, M.R.; Johnson, R.L.; Pankow, J.F. *Envir. Sci. Technol.* **1992**, *26*, 901-908.
(11) Hunt, J.R.; Sitar, N.; Udell, K. *Water Res. Res.* **1988**, *24* 1247-1258.
(12) Mackay, D.M.; Cherry, J.A. *Env. Sci. Tech.* **1989**, *23* 620-636.
(13) West, C.C.; Harwell, J.H. *Environ. Sci. Technol.* **1992**, *26*, 2324-2330.
(14) Pickens, J.F.; Jackson. R.E.; Statham, W.H.; Brown, C.L.; Pope, G.A. *in HAZMAT Southwest Conference Proceedings, September 28-30, Dallas, Texas,* **1993**.
(15) Powers, S.E.; Abriola, L.M.; Weber, W.J. *Water Res. Res.* **1994**, *30*, 321-332.
(16) Abdul, A.S.; Gibson, T.L.; Rai, D.N. *Groundwater.* **1990**, *28*, 920-926.
(17) Fountain, J.C.; Klimek, A.; Beikirch, M.; Middleton, T. *J. Haz. Mater.* **1991**, *28*, 295 - 311.
(18) Wunderlich, R.W.; Fountain, J.C.; Jackson, R.E. *J. Soil Contam.* **1992**, *1*, 361-378.
(19) Johnson, R.L.; Pankow, J.F. *Environ. Sci. and Technol.* **1992**, *26*, 896-901.
(20) Rouse, J.D.; Sabatini, D.A. *Environ. Sci. Technol.* **1993**, *27*, 2072-2078.
(21) Pennell, K.D.; Abriola, L.M.; Weber, W.J. *Envir. Sci. Technol.* **1993**, *27*, 2332-2340.
(22) Kann, A.T.; Tomson, M.B. *Env. Tox. Chem.* **1990**, *9*, 253-263.

(23) Aronstein, B.N.; Calvillo, Y.M.; Alexander, M. *Soil Env. Sci. Technol.* **1991**, *25*, 1728-1731.

(24) Lake, L.W. *Enhanced Oil Recovery*, Prentice Hall: Englewood Cliffs, NJ, 1989.

(25) Wilson, J.L.; Conrad, S.H., In *Proceedings of NWWA/API Conference On Petroleum Hydrocarbons and Organic Chemicals in Groundwater, Nov. 1984, Houston Texas*; National Water Well Association: 1984; pp. 274-298.

(26) Parker, J.C. *Rev. Geophysics.* **1989**, *27*, 311-328.

(27) Mayer, A.S.; Miller, C.T. *J. Cont. Hydr.* **1992**, *11*, 189-213.

(28) Geller, J.T.; Hunt, J.R. *Water Res. Res.* **1993**, *29*, 833-845.

(29) Chiu, C.T.; Porter, P.E.; Schmedding, D.W. *Envir. Sci. Technol.* **1983**, *17*, 227-321.

(30) Hoffman, F. *Groundwater.* **1993**, *31*, 98-106.

(31) Scamehorn, J.F.; Schechter, R.S.; Wade, W.H. *J. Colloid and Interface Sci.* **1982**, *85*, 463-501.

(32) von Rybinski, W.; Schwuger, M.J. In *Nonionic Surfactants: Physical Chemistry*; M.J. Schick, Ed.; Surfactant Science Series; Marcel Dekker Inc.: New York, NY, 1987, Vol. 23.

(33) Allred, B; Brown, G.O. *GWMR.* 1994, *Spring*, 174-184.

(34) *Nonionic Surfactants: Physical Chemistry*, M.J. Schick, Ed.; Surfactant Science Series; Marcel Dekker Inc.: New York, NY,1987; Vol. 23.

(35) Grim, R.E. *Clay Mineralogy*, 2nd Edition, McGraw-Hill Book Co.: New York, NY, 1968; Chapter 7.

(36) Swisher, R.D. *Surfactant Biodegradation*, 2nd edition, Marcel Dekker, Inc.: New York, NY, 1987.

(37) Federle, T.W; Schwab, B.S. *Water Res.* **1992**, *26*, 123-127.

(38) Ahel, M.; Giger, W.; Koch, M. *Water Res.* **1994**, *28(5)*, 1131-1142.

(39) Fountain, J.C. In *Transport and Remediation of Subsurface Contaminants*; Sabatini, D.A.; Knox, R.C., Eds.; American Chemical Society Symposium Series; ACS: Washington, DC., 1992, Vol. 491, Chapter 15.

(40) Palmer, C.D.; Fish, W. *"Chemical Enhancements to Pump and Treat Remediation*, EPA/540/S-92/001; Office of Research and Development, U.S. EPA: Washington, DC, 1992.

(41) Kueper, B.H.; Frind, E.O. *Water Res. Res.* **1991**, *27*, 1049-1057.

(42) Kueper, B.H.; McWhorter, D.B. *Groundwater.* **1991**, *29*, 716-728.

(43) Rosen, M.J., *Surfactants and Interfacial Phenomena*, 2nd edition, John Wiley & Son: New York, NY, 1989.

(44) Desmond, A.H.; Desai, F.N.; Hayes, K.F.; *Water Res. Res.* **1994**, *30*, 333-342.

(45) Jackson, R.E.; Pickens, J.F.; Huag, A. in *HAZMAT Southwest Conference Proceedings, September 28-30, Dallas, Texas*, **1993**.

RECEIVED January 24, 1995

Chapter 14

Modeling the Surfactant-Enhanced Remediation of Perchloroethylene at the Borden Test Site Using the UTCHEM Compositional Simulator

G. A. Freeze[1], J. C. Fountain[2], G. A. Pope[3], and R. E. Jackson[4]

[1]INTERA Inc., 1650 University Boulevard, Suite 300, Albuquerque, NM 87102
[2]Department of Geology, State University of New York, Buffalo, NY 14260
[3]Department of Petroleum Engineering, University of Texas, Austin, TX 78712
[4]INTERA Inc., 6850 Austin Center Boulevard, Suite 300, Austin, TX 78731

The UTCHEM multiphase compositional simulator was used to model the surfactant-enhanced remediation of perchloroethylene (PCE) in a 3 meter by 3 meter test cell at Canadian Forces Base Borden in Alliston, Ontario. A total of 231 liters of PCE was injected into the center of the test cell. After 27 days, time domain reflectometry (TDR) measurements indicated that PCE migration in the test cell was essentially complete. An aqueous surfactant solution was then circulated through the test cell via a system of injection and withdrawal wells to recover the injected PCE.

UTCHEM is a three-dimensional, multiphase, multicomponent, compositional simulator capable of modeling fluid flow and mass transport in aquifers undergoing remediation. A vertically heterogeneous layered model was created with physical properties estimated from field measurements. Surfactant and phase transition properties were derived from laboratory data. 201 days of surfactant flooding were simulated, during which 14.4 pore volumes of aqueous surfactant solution had been circulated (at rates of between 0 and 600 gpd) removing approximately 60% of the PCE. UTCHEM was able to closely reproduce the PCE recovery over time and the PCE distribution after 201 days of surfactant flooding.

The favorable comparison of UTCHEM results with field test results demonstrates the utility of UTCHEM in predicting surfactant-enhanced remediation processes. UTCHEM can be used both for site characterization and as a model to test surfactant effectiveness and compare remediation options.

0097–6156/95/0594–0191$12.00/0

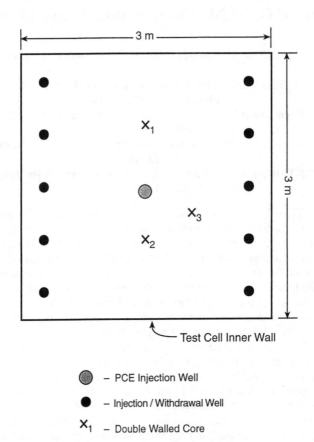

Figure 1. Plan view of test cell at Borden test site.

The controlled release and subsequent migration of perchloroethylene (PCE) in a 3 m by a 3 m test cell at Canadian Forces Base Borden near Alliston, Ontario has been previously documented (*1*). The test cell (Figure 1) was created by driving sheet piling through 4 m of water-saturated layered sand and into an underlying clay aquitard. A line of five injection wells was installed on one side of the test cell and a line of five withdrawal wells was installed on the opposite side of the cell. The injection and withdrawal wells penetrated the entire depth of the sand aquifer. A shallow well was utilized to inject 231 liters of PCE into the center of the test cell. PCE injection started on June 12, 1990 and continued for about 29 hrs at a relatively constant rate of about 8 liters/hr. Time domain reflectometry (TDR) measurements suggested that the migration of PCE within the test cell was essentially complete 27 days after the end of the injection period.

On July 11, 1990, the upper 1 m of the saturated sand was excavated and replaced with a confining bentonite layer. Based on observations from the excavation (*1*), it was determined that the PCE migration followed the horizontal bedding of the sand and that it migrated preferentially through the coarser grained sand units. PCE saturations were observed to be highest near the center of cell although PCE had reached the cell walls in the coarser grained layers. A total of 52 liters of PCE was present in the excavated sand. In August, 1990, three cores were taken from the lower portion of the test cell. There was reasonable correlation of PCE saturation with depth between the cores. The saturation distribution from Core 3, located near the center of the test cell, is shown in Figure 2. Preferential migration is evidenced by differences in PCE saturations with depth. The maximum residual PCE saturation is about 0.15 (*1*). Free-phase PCE was observed during excavation (*1*) and is suggested by PCE saturations near and above 0.15 in Core 3.

The extraction of PCE from the test cell has been described in detail elsewhere (*2*). The remediation process involved (i) direct pumping of free-phase PCE, (ii) water flooding to remove free-phase and dissolved PCE, and (iii) surfactant flushing to solubilize additional residual PCE. Direct pumping of PCE from the wells was performed for about 2 weeks, during which 47 liters of PCE were recovered. Water flooding (pump-and-treat), using the injection-withdrawal well system, took place throughout October, 1990, yielding an additional 12 liters of PCE. An aqueous surfactant solution of 1% (by weight) nonyl phenol ethoxylate (NP 100) and 1% (by weight) phosphate ester of the nonyl phenol ethoxylate (Rexophos 25-97) was then circulated through the test cell via the injection-withdrawal wells. Between November 11, 1990 and May 29, 1991, a total of 130,000 liters (14.4 pore volumes) of surfactant solution were recirculated through the test cell, during which time 62 liters of PCE were recovered. PCE was removed from the effluent prior to reinjection. Pumping was intermittent (Figure 3), with maximum rates of about 2,300 liters/day (600 gallons/day).

The UTCHEM multiphase compositional simulator was used to model the surfactant flushing phase only. This modeling study demonstrates the capability of UTCHEM compositional simulator to model the surfactant-enhanced remediation of DNAPL at a field site.

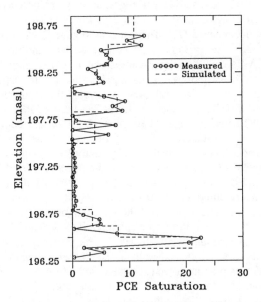

Figure 2.   Measured and simulated PCE saturation at the location of Core 3 prior to surfactant flooding.

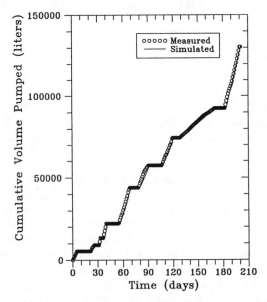

Figure 3.   Measured and simulated pumping rates during surfactant flushing period.

Model Conceptualization

UTCHEM is a three-dimensional, multiphase, multicomponent, compositional simulator capable of modeling flow and mass transportation in aquifers undergoing remediation. The capabilities of UTCHEM are well documented (*3, 4, 5*). UTCHEM can model up to four phases (aqueous, gas, NAPL, microemulsion), up to 18 components (including water, PCE, and surfactant), and has the capability to simulate enhanced solubilization and increased mobilization resulting from surfactant injection. For this modeling study, the aqueous phase represented water with dissolved PCE, the NAPL phase was free phase PCE, and the microemulsion phase was an aqueous solution containing water, surfactant, and dissolved PCE. Only the 201-day surfactant flushing period was simulated.

Prior to surfactant flooding, at least 111 liters of PCE had been removed from the test cell as a result of excavation (52 liters), free-phase pumping (47 liters), and water flooding (12 liters). The remaining 120 liters (from the 231 liters initially injected) represents an upper bound on the initial "pre-surfactant" PCE volume present in the test cell. Some PCE may also have been lost due to volatilization from the surface of the cell (*2*), in which case the "pre-surfactant" PCE volume would be less than 120 liters. Continued surfactant flushing in late 1991 and early 1992 produced an additional 17 liters of PCE from the test cell beyond the 62 liters produced during the simulated 201-day recovery period. Therefore, prior to surfactant flooding at least 79 liters of PCE was available for remediation in the test cell.

A three-dimensional UTCHEM grid was created. Horizontal discretization (Figure 4) represented a half-cell with centerline symmetry assumed. Vertical heterogeneity was incorporated by discretizing 14 layers. Simulated initial PCE saturations for each of the layers at the center of the test cell are shown in Figure 2. Simulated initial saturations in each layer were decreased with distance away from the center of the cell, to be consistent with observations in the excavated portion of the test cell (*1*). The assumed initial distribution corresponds to an initial PCE volume of 105 liters.

Physical properties of the Borden sand have been extensively measured (*6, 7, 8*). UTCHEM input parameters were selected to be consistent with these measured values. A porosity of 0.39 was simulated, corresponding to a pore volume of about 9,000 liters (approximately 2,400 gallons). Simulations used 0.03 m for longitudinal dispersivity and 0.01 m for transverse dispersivity. Relative permeability and capillary pressure relationships were specified to reproduce measured data (*9, 10*). The methodology is described in (*11*). Laboratory permeameter tests of test cell cores were used to determine hydraulic conductivity variations with depth. In creating the vertical discretization, an attempt was made to preserve the observed conditions where layers with relatively high PCE saturations are underlain by low permeability layers. The simulated hydraulic conductivities in each of the 14 vertical layers is shown in Figure 5. Note that the hydraulic conductivity varies only by about a factor of 3, from approximately 0.003 to 0.010 cm/s. Residual saturations were 0.17 for PCE and 0.31 for water. Laboratory experiments showed the surfactant-enhanced solubility of PCE to be about 11,700 ppm as compared to an aqueous solubility of about 200 ppm. The injection of the surfactant solution in UTCHEM resulted in the conversion of the aqueous phase to a microemulsion phase. The microemulsion phase was specified

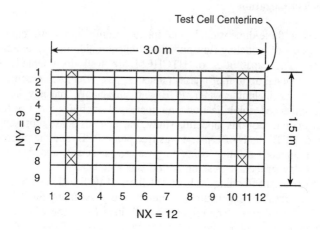

X – Injection / Withdrawal Well

Figure 4.   Horizontal discretization of one half of Borden test cell (no flow across cell centerline is assumed).

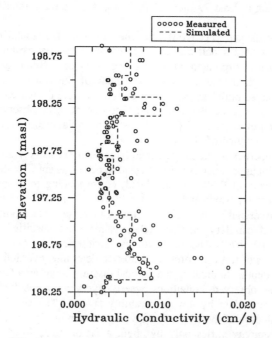

Figure 5. Measured and simulated hydraulic conductivity as a function of depth.

to have properties similar to water but with the surfactant-enhanced solubility (11,700 ppm).

Model Results

The 201 day simulation of surfactant flooding required 4 hrs on a 66 MHz 486-based PC. Simulation results were compared to measured field test results for PCE recovery (Figure 6), effluent concentration (Figure 7), and PCE saturation. The general trends from the field test results were reproduced quite well by the UTCHEM simulation results. Simulated PCE recovery (Figure 6) was less than the measured recovery at early time and greater than the measured recovery at later times. The periods with no PCE recovery correspond to pump downtime. The simulated effluent concentration (Figure 7) peaked at early time at lower than the measured value but was greater than the measured effluent concentration at later times. These observations about effluent concentrations are consistent with the PCE recovery behavior. Both the measured and simulated final PCE saturations showed that most of the PCE remained in a pool at the bottom of the test cell.

At early time (the first one or two pore volumes), PCE removal is controlled by the volume of free-phase PCE present (i.e., PCE at saturations greater than the residual saturation of 0.17) and by removal from the more transmissive layers. The simulation results indicate lower than measured early time PCE removal, which suggests that the initial volume of free-phase PCE present was larger than simulated and/or that there are some layers with higher-than-simulated hydraulic conductivity controlling early-time PCE removal. In the UTCHEM simulation, 75 liters of PCE were removed over the 201 days of surfactant flooding as compared with a measured value of 62 liters removed (Figure 6). This result suggests that the total volume (free-phase plus dissolved) of PCE present prior to surfactant flooding may have been less than 105 liters and/or that some of the remaining PCE was in layers with lower-than-simulated hydraulic conductivity. Excavation of the test cell following surfactant flooding located at least 9 liters of PCE trapped in indentions in the underlying clay aquitard, suggesting that volume of PCE available for remediation may have been less than simulated.

The effects of the intermittent pumping rate are evidenced by sharp spikes in both the measured and simulated effluent concentrations (Figure 7). Pump downtime increases the in-situ residence time of the surfactant solution, which increases the amount of PCE solubilized, and results in a delayed step increase in effluent concentration when the pump is turned back on. Pump downtime also leads to surfactant decay (which decreases effluent concentration) and in-situ biological activity (which increases effluent concentration). Surfactant decay and biological effects were not simulated and could be partially responsible for differences between simulated and measured PCE removal.

Conclusions

The general trends from the Borden field test results (PCE recovery, effluent concentration, and final PCE saturation) for surfactant-enhanced remediation of PCE were reproduced using UTCHEM. Differences between measured and simulated results were attributed to uncertainties in the initial PCE volume and distribution and in the hydraulic conductivities within the layers. Parameter

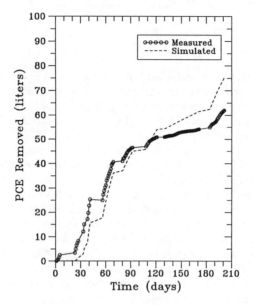

Figure 6.   Measured and simulated volume of PCE removed from the test cell during surfactant flooding.

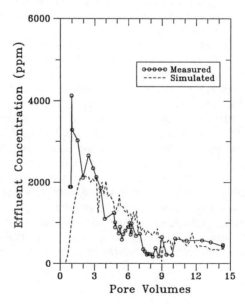

Figure 7.   Measured and simulated PCE concentration in effluent produced during surfactant flooding.

sensitivity simulations could resolve some of the uncertainty, but were not performed because a favorable comparison of UTCHEM results with field test results was obtained using simplified layering and best estimates of properties. Additional simulations of the Borden test site could demonstrate sensitivity to variations in physical properties, surfactant properties, initial conditions, and pumping schemes.

The reproducibility of the remediation sequence with a single deterministic simulation demonstrates the utility of UTCHEM in predicting surfactant-enhanced remediation processes on a macroscopic scale. A more detailed approximation of the fine vertical layering, perhaps in a stochastic framework, is necessary to reproduce PCE migration because of the lower gradients (i.e., no pumping), but was not a part of this modeling study. UTCHEM can be used for (i) inverse determination of hydrogeological characterization of aquifer properties and DNAPL distribution, and (ii) predictive modeling to examine remediation alternatives, optimal surfactant properties (solibility, mobility), pumping schemes, and DNAPL recovery. Parameter sensitivity simulations can identify which parameters have a significant impact on simulation results. The most sensitive parameters must be well defined from field or laboratory tests for the predictive capabilities of UTCHEM to be fully utilized.

Literature Cited

(*1*)   Kueper, B.H. D. Redman, R.C. Starr, S. Reitsma, and M. Mah, A Field Experiment of Study the Behavior of Tetrachloroethylene Below the Water Table; Spatial Distribution of Residual and Pooled DNAPL, *GROUND WATER*, Vol. 31, No. 5, September - October 1993.

(2)   Fountain, J.C. and D.S. Hodge, Extraction of Organic Pollutants Using Enhanced Surfactant Flushing - Initial Field Test (Part 1), Project Summary, New York State Center for Hazardous Waste Management, State University of New York at Buffalo, February, 1992.

(3)   Datta Gupta, A., G.A. Pope, K. Sepehrnoori, and R.L. Thrasher, A Symmetric, Positive Definite Formulation of a Three-Dimensional Micellar/Polymer Simulator, *SPE Reservoir Engineering*, 1(6), 622, 1986.

(4)   Bhuyan, D.; Pope, G.A.; Lake L.W. Mathematical Modeling of High-pH Chemical Flooding, SPE Reservoir Engineering, 5(2), 213, 1990.

(5)   Saad, N., G.A. Pope, and K Sepehrnoori, Application of Higher-Order Methods in Compositional Simulation, *SPE Reservoir Engineering*, (5)4, 623, 1990.

(6)   MacFarlane, D.S., J.A. Cherry, R.W. Gillham, and E.A. Sudicky, Migration of Contaminants in Groundwater at a Landfill: A Case Study , 1. Groundwater Flow and Plume Delineation, J. of Hydrology, Vol, 63, 1-29, 1983.

(7)   Mackay, D.M., D.L. Freyberg, and P.V. Roberts, A Natural Gradient Experiment on Solute Transport in a Sand Aquifer. 1. Approach and Overview of Plume Movement, *Water Resources Research*, Vol. 22, No. 13, 2017-2029, 1986.

(8)   Sudicky, E.A., A Natural Gradient Experiment on Solute Transport in a Sand
      Aquifer: Spatial variability of Hydraulic Conductivity and Its Role in the
      Dispersion Process, *Water Resources Research*, Vol. 22, No. 13, 2069-2081,
      1986.

(9)   Kueper, B.H., The Behavior of Dense, Non-Aqueous Phase Liquid
      Contaminants in Heterogeneous Porous Media, Ph.D. dissertation, The
      University of Waterloo, Ontario, 1989.

(10) Kueper, B.H. and E.O. Frind, Two-phase Flow in Heterogeneous Porous
      Media. 2. Model Application, *Water Resources Research*, Vol. 27, No. 6,
      1059-1070, 1991.

(11) Brown, C.L., G.A. Pope, L.M. Abriola, and K. Sepehrnoori, Simulation of
      Surfactant Enhanced Aquifer Remediation, Draft submitted to WRR, June,
      1993.

RECEIVED December 13, 1994

Chapter 15

An Interwell Solubilization Test for Characterization of Nonaqueous-Phase Liquid Zones

George W. Butler[1], Richard E. Jackson[1], John F. Pickens[1], and Gary A. Pope[2]

[1]INTERA Inc., 6850 Austin Center Boulevard, Suite 300, Austin, TX 78731
[2]Department of Petroleum Engineering, University of Texas, Austin, TX 78712

It is essential that the location and chemical composition of non-aqueous phase liquids (NAPLs) be characterized for their timely and cost-effective remediation. Approaches using ground-water sampling and core evaluations do not provide adequate characterization capabilities. The use of surfactant solutions in single- and interwell NAPL solubilization tests (NSTs) provides a means to locate and characterize NAPLs in the subsurface. By injection and recovery of micellar surfactant or cosolvent solutions, significant volumes of an aquifer can be tested for the presence of a NAPL. Positive indications of NAPL can be determined based on the recovery of dissolved NAPL components at concentrations above their aqueous solubilities or based on significant increases in dissolved component recoveries associated with the test. Owing to a surfactant's capability to significantly increase the recovery of low solubility hydrophobic components of the NAPL, the complete chemical composition of the NAPL can be determined with the NAPL solubilization test, in contrast to routine ground-water monitoring surveys where only the more soluble components are identified. Two-dimensional, cross-sectional numerical simulations show that interwell NSTs can be used to positively identify the presence of a NAPL. The simulations show that a surfactant solution sweep in the presence of a NAPL causes a significant increase in NAPL recovery (including recoveries greater than the aqueous solubility of the NAPL) relative to conventional ground-water extraction. The increase in NAPL recovery is evident, even with the effects of dilution due to the extraction of uncontaminated waters.

The characterization of non-aqueous phase liquids (NAPLs) in the subsurface poses much difficulty for hydrogeologists. In their review of the effects of NAPLs on pump-and-treat remediation, Mackay and Cherry *(1)* noted that "very little success has been achieved in even locating the subsurface (NAPL) sources, let alone

0097–6156/95/0594–0201$12.00/0
© 1995 American Chemical Society

removing them." NAPL site-characterization methods generally involve the recovery of core from the contaminated aquifer and the chemical analysis of subsections of core. Recent developments allow the interpretation of these analytical data to provide estimates of the likelihood of the presence of NAPL on the basis of partitioning calculations (2) or the use of dyes added to a liquid extract of the core to indicate visually the presence of NAPL (3).

However, limitations to these approaches have been demonstrated by the work by Mayer and Miller (4), which has shown that the scale of measurement for residual NAPL probably is much larger than the aquifer sample provided by a core. Furthermore, they noted that as the porous medium becomes more non-uniform, the necessary volume of aquifer to be sampled to yield a representative value of residual NAPL saturation (percentage of pore volume occupied by NAPL) increases rapidly. Therefore, methods are required to sample larger volumes of aquifer than can be tested using cores from boreholes. Thus, well-test methods appear to be particularly attractive for successful characterization of the subsurface distribution of NAPLs. By analogy, hydrogeologists have long used pumping tests to measure the hydraulic conductivity field around a pumping well, rather than relying on permeameter measurements of cores taken during well installation.

The purpose of this paper is, first, to demonstrate by numerical simulation how well-test methods involving the injection and extraction of micellar surfactant solutions can be used to detect NAPL zones in the subsurface. This method has been developed by Jackson and Pickens (5) to address the critical need for improved methods for characterization of sites contaminated by NAPLs. Second, we discuss the solubilization principles by which surfactant solutions might be formulated to detect very low solubility components of NAPLs.

Characterization of NAPL Zones

The concern with locating NAPL zones has led the EPA (6) to recommend the following guidance on the characterization of all hazardous waste sites for NAPLs.

1. *The likelihood of subsurface NAPL contamination should be evaluated as a part of all site investigations.*

2. *If NAPL contamination is likely, characterization of the potential nature and extent of such contamination is recommended to determine appropriate remedial actions.*

Quite clearly, if remedial technologies are to be focused on NAPL zones, it is essential that the NAPL zones be properly located and mapped with respect to high- and low-permeability stratigraphic units and permeability discontinuities within these units. The lack of useful, invasive NAPL site-characterization techniques has prompted EPA (7) to call for "improved field methods for rapid and inexpensive detection of NAPL that cannot be readily identified by visual inspection of drilling samples."

Not only is the location of the NAPL zone required before effective aquifer remediation might begin, the quantitative chemical analysis of the NAPL is also required because it is entirely possible that only the more soluble components of a NAPL pool may have dissolved and been observed at nearby monitoring wells, a matter identified over 10 years ago *(8)* but largely ignored by hydrogeologists. This issue of the seemingly invisible, low-solubility NAPL components becomes critical if the NAPL zone contains such components that require special treatment according to applicable regulations, e.g., PCBs and chlorinated pesticides, rather than simple air stripping which is suitable for volatile organic chemicals. Thus, the characterization of NAPL zones should involve field methods that can effectively sample and identify all components of the NAPL.

A third and final requirement for the effective characterization of NAPL zones is the quantitative measure of the fraction of the total pore space occupied by the NAPL, i.e., the residual NAPL saturation (S_r). This measure must be obtained if meaningful estimates are to be made of the NAPL mass to be recovered and treated and if quantitative performance assessments are to be conducted to monitor the progress of remedial operations. Such performance assessments conducted during a site remediation are essential to providing meaningful estimates of remediation times and costs to complete a site cleanup. We have shown how the average residual saturation of a NAPL zone and the volume of NAPL it contains may be estimated by the application of a NAPL partitioning-tracer test or NPTT *(9, 10)*.

Therefore, we identify three goals in characterizing NAPL zones: (i) the detection of the location of a NAPL zone in a test section of an aquifer, (ii) the identification of its chemical composition, and (iii) the measurement of the volume of the NAPL in the test section. A combination of well-test methods using NAPL micellar-solubilization and partitioning-tracer tests can be used to meet these goals. In this paper, we present simulations to demonstrate the principles of the micellar-solubilization tests.

The Principle of Napl Solubilization Testing

Because surfactants are capable of the *in situ* solubilization of a wide variety of organic compounds, injection-extraction well-test methods involving surfactant solutions are ideal for locating and sampling NAPLs. This principle has been used *(5)* to develop an *in situ* method for locating and compositionally characterizing NAPLs. This method may employ either a single injection-extraction well or an injection-extraction well pair, or even several wells with one or more injectors and one or more producers.

Fountain *(11;* also Fountain et al., *Ground Water,* in press; and *12)* has shown in field experiments in Canada and Texas that dense NAPLs (DNAPLs), such as perchloroethylene and carbon tetrachloride, can be successfully removed from sandy aquifers by flushing the contaminated aquifer with surfactant solutions. These chlorinated solvents were solubilized by nonionic and anionic surfactants such that *in situ* concentrations of the chlorinated solvents measured in monitoring wells were more than an order of magnitude greater than their aqueous solubilities.

Figure 1 shows the enhanced solubility of PCE in the effluent measured from the Borden subsurface cell experiment *(11)* due to its solubilization by the injection of a 2% micellar surfactant solution which almost completely decontaminated the cell of tetrachloroethene (PCE) NAPL *(13)*. Jackson et al. *(14)* used this characteristic in their discussion of single- and interwell NAPL solubilization tests to locate and chemically characterize NAPL zones in sand and gravel aquifers, as well as to remove NAPLs from the subsurface after their detection.

For surfactant concentration above the critical micelle concentration or CMC *(15, 16)*, the effective solubilities of NAPL chemicals can be increased one or more orders of magnitude above the aqueous solubility of that chemical. Micelles are colloidal-sized aggregates of surfactants, the interior of which provide a hydrophobic environment in which NAPLs can be solubilized. The solubility enhancements created by injecting micellar surfactant solutions into the subsurface are sufficient to raise the measured concentrations of solubilized NAPL in the surfactant solutions well above the aqueous solubilities of the particular NAPL contaminants, even allowing for much dilution of the measured concentration by a variety of processes (Jackson and Mariner, *Ground Water,* in press).

The detection of low-solubility NAPL components within a NAPL mixture containing higher solubility components, e.g., trace quantities of PCBs within a chlorinated solvent NAPL, can be accomplished by the micellar solubilization of the NAPL. It has been shown for both polycyclic aromatic hydrocarbons *(17)* and chlorinated hydrocarbons *(12)* that the lower the aqueous solubility of a particular hydrocarbon, the greater its enhancement by micellar solubilization. Thus PCE, with an aqueous solubility of 240 mg/L, can have its effective solubility in a 1% micellar surfactant solution raised one hundredfold, while 1,2-dichloroethane (aqueous solubility = 8690 mg/L) can be enhanced only two- to threefold *(12)*. That is, the micelle-NAPL partition coefficient for PCE is much greater than that for 1,2-dichloroethane.

By the appropriate choice of surfactants, it is possible to selectively enhance the solubilization of low-solubility components in a NAPL mixture and therefore detect trace quantities of this low-solubility component. For example, Fountain et al. *(12)* reported on the selective extraction of trace PCBs from a transformer oil using a surfactant that was an optimal solubilizer for PCBs. The choice of surfactant to detect low-solubility components must be based on the selectivity ratio for the trace compound ("A") which is codissolved in a bulk phase ("B"), e.g., trace PCBs in a chlorinated solvent such as PCE.

Nagarajan and Ruckenstein *(18)* have defined the selectivity ratio for A in a binary mixture of A and B as:

$$SR^A = \frac{\left(X^A_{mic}/X^B_{mic}\right)}{\left(X^A_{NAPL}/X^B_{NAPL}\right)} \tag{1}$$

where X^A_{mic} and X^B_{mic} are the mole fractions of A and B in the micellar solution, and X^A_{NAPL} and X^B_{NAPL} are the mole fractions of A and B in the NAPL. This equation can be rewritten to show that the selectivity ratio is also the ratio of the micellar-NAPL partition coefficients. Since the micellar-aqueous phase partition coefficient for the solubilized contaminant A is given by Edwards et al. *(17)* as $K^A_{mic-aq} = X^A_{mic}/X^A_{aq}$, where X^A_{aq} is the aqueous phase concentration of A, and because X^A_{aq} is defined as the product of the solubility of A in a NAPL-saturated solution at the CMC ($S^A_{NAPL,CMC}$) and the molar volume of water, we may rewrite equation 1 as:

$$SR^A = \frac{\left(K^A_{mic-aq} S^A_{NAPL,CMC}\right)}{\left(K^B_{mic-aq} S^B_{NAPL,CMC}\right)} \cdot \frac{\left(X^B_{NAPL}\right)}{\left(X^A_{NAPL}\right)} \tag{2}$$

The right-hand side of equation 2 indicates that the selectivity ratio of the solubilized contaminant A, SR^A, is a function of two terms. The first of these terms is dependent on the properties of the micellar surfactant solution which dictates the values of the micellar-aqueous partition coefficient for A and the solubility of the NAPL component A at the CMC. In order to detect low-solubility components, a micellar surfactant solution would be chosen to maximize the value of this first term. The second term is a simple function of the composition of the NAPL ($X^A << X^B$) to be sampled during the NAPL solubilization test.

The Single-Well NAPL Solubilization Test

The single-well NAPL solubilization test (NST) involves injecting a dilute surfactant or cosolvent solution into a contaminated aquifer through a well, allowing the solution to travel radially from the well some radius and then back producing the solution through the well. While in the subsurface, the solution will solubilize NAPLs, if present, and, upon back production, transport the solubilized NAPL components to the well. The radial distance to which the injected surfactant solution is greater than the critical micelle concentration is called the critical radius. By employing several single-well tests with overlapping critical radii, the zone of NAPL residuals can be located with increasing precision.

The single-well test can also be utilized to estimate the mass fractions of each of the components of the NAPL zone so that the above-ground treatment operations can be designed to allow for extraction of low-solubility components of the NAPL not previously identified in the ground-water monitoring surveys. A further use of the single-well test is the evaluation of the remediation efficacy under *in situ* conditions of a particular surfactant or cosolvent solution, chosen on the basis of the complete chemical analysis of the NAPL, prior to implementation of full-scale remediation. We have illustrated the application of the single-well NAPL solubilization test elsewhere *(14)*.

The test can equally well involve different injection and extraction wells, a methodology known as the interwell NST *(19)*. There are several advantages to using interwell tests rather than single-well tests. Most importantly, they offer the ability to sample various parts of a contaminated aquifer simultaneously rather than sequentially, as is the case with the single-well test and the potential for using the

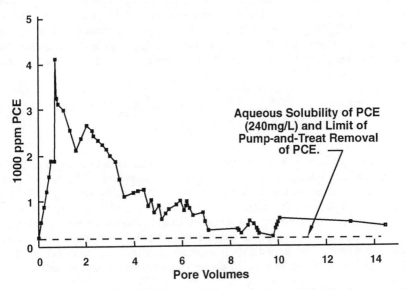

Figure 1. Effluent concentration of perchloroethylene (PCE) from Borden cell showing the enhancement in PCE solubility due to micellar solubilization by surfactants (adapted from Fountain, 1992).

1 DRUM PCE SPILL

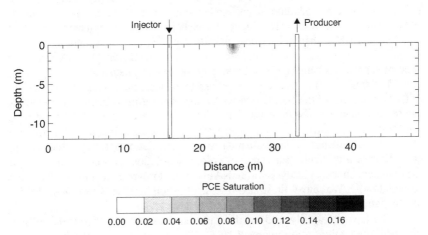

Figure 2. Cross-sectional model showing the 1-drum PCE spill as ganglia.

interwell configuration in the subsequent remedial operations. An application involving an interwell NST follows.

Numerical Simulation of an Interwell NAPL Solubilization Test

A two-dimensional cross-sectional model was used to simulate three interwell NAPL solubilization test cases. UTCHEM (the University of Texas Chemical Flood Simulator, *(20)*; Delshad et al., unpublished data), a three-dimensional, multi-phase, multi-component, finite-difference, numerical simulator was used for the interwell tests. Simulations were performed for one- and five-drum PCE spills where the spilled PCE was trapped as ganglia and for a five-drum PCE pool at the base of the cross section (1 drum = 55 U.S. gallons = 208 L). The cross-sectional model consists of a single confined sand aquifer with stochastically variable hydraulic conductivity. An upstream injection well and a downstream extraction well are located around a central spill area. The model conceptualization was initially developed by Brown *(21* and Brown et al. *(22)* using the geologic characteristics of a sand aquifer at the Canadian Forces Borden site. The model configuration was modified for the interwell NAPL solubilization test simulations presented below.

The model parameters for the aquifer, PCE, and surfactant solution are summarized in Table I. The aqueous solubility of PCE used in the simulations is 240 mg/L. The surfactant used is a 50:50 mixture of sodium diamyl and dioctyl sulfosuccinates (AY-OT). A 1% surfactant solution was used in the NAPL solubilization test simulations. The following surfactant solution properties are based on laboratory work performed by Minquan Jin of the University of Texas at Austin. A 1% AY-OT surfactant solution will have an interfacial tension with PCE of approximately 1 dyne/cm. The solubilization ratio, or volume of PCE which can be dissolved per volume of surfactant in the surfactant solution, is 0.5 for a 1% surfactant solution. This results in a solubility of PCE of approximately 8,100 mg/L of PCE in 1% surfactant solution. The sorption of surfactant is 2.4 mg/g of soil. Sorption of the PCE was not incorporated in the simulations.

The results of the simulations are based on pore volumes extracted from the system where one pore volume is the volume of fluid within the cross section between the injection and extraction wells and within the screened intervals of the wells. In all the simulations, the injection and extraction volumes were balanced. Initially, two pore volumes of ground water were injected and extracted to simulate the effects of balanced pump-and-treat remediation. This was followed by injection of one pore volume of 1% surfactant solution. Several pore volumes of ground water were injected and extracted to flush the surfactant from the system.

Figures 2 and 3 show the cross section with the one- and five-drum PCE spills, respectively, as ganglia. That is, the PCE is distributed in the sand aquifer as residual DNAPL held by capillary forces. The screened intervals of both the injector and extractor run from the top to the bottom of the cross section in both cases. Figure 4 shows the results of the interwell simulations for the one- and five-drum ganglia. The results show that the PCE recovery during the initial pump-and-treat portion of the test is at concentrations less than the aqueous

Table I. Interwell NAPL Solubilization Test Model Parameters

MODEL PARAMETER	VALUE
Grid dimensions	49 m long, 1 m wide, and 12 m thick (160.8 ft x 3.28 ft x 39.4 ft)
Grid blocks	29 x 24
Porosity	0.34
Hydraulic conductivity range	8×10^{-6} to 4×10^{-4} m/s
Mean hydraulic conductivity	8×10^{-5} m/s
K_v/K_h	0.5
Hydraulic gradient	0.0043 m/m
Longitudinal dispersivity	0.03 m
Transverse dispersivity	0.01 m
Surfactant type	50:50 mixture of sodium diamyl and dioctyl sulfosuccinates
Surfactant concentration	1%
Surfactant CMC (critical micelle concentration)	0.0001 (volume fraction)
Density	Water: 1.00 g/cc PCE: 1.625 g/cc Surfactant: 1.15 g/cc
Viscosity	water: 1 cp PCE: 0.89 cp
Interfacial Tension	45 dyne/cm (PCE/water) 1 dyne/cm (PCE/1% surfactant solution)
PCE aqueous solubility	240 mg/L
Solubilization ratio, V_{PCE}/V_{surf}	0.5 (1% surfactant solution)
Surfactant sorption	2.4 mg/g of soil
Residual saturations	Water: 0.24 PCE: 0.17

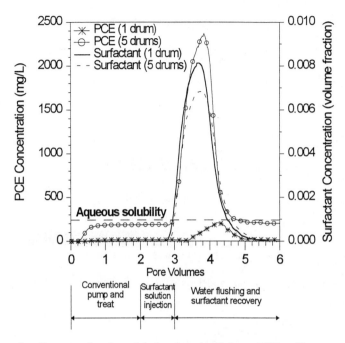

Figure 3. Cross-sectional model showing the 5-drum PCE spill as ganglia.

5 DRUM PCE SPILL

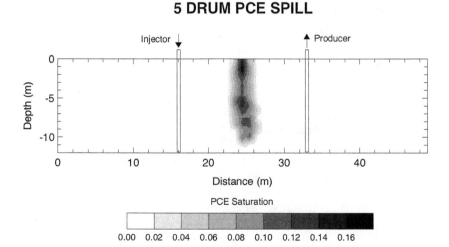

Figure 4. Comparison of results of interwell NAPL solubilization tests for the 1- and 5-drum PCE ganglia cases.

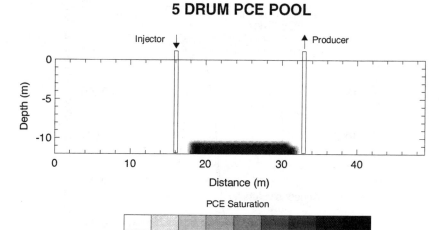

Figure 5.   Cross-sectional model showing the 5-drum PCE spill as a pool.

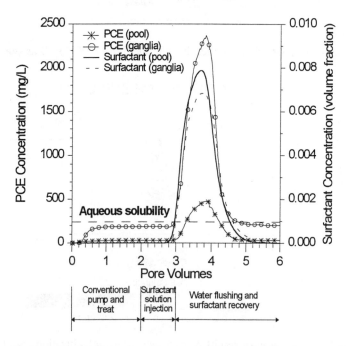

Figure 6.   Comparison of results of interwell NAPL solubilization tests for the 5-drum ganglia and 5-drum pool cases.

solubility of PCE due to dilution by mixing of contaminated and uncontaminated water entering the extraction well. However, once the NAPL solubilization test is under way and the surfactant solution containing solubilized PCE reaches the extraction well, the PCE recovery increases dramatically. After the pore volume of surfactant solution has swept through the system, the PCE concentrations drop back to pump-and-treat ranges. The difference in the PCE concentrations recovered in the one- and five-drum spills is due to the greater dilution which occurs in the one-drum case because of reduced area for contact between the PCE and surfactant solution.

Four points can be made to show that, in addition to a dissolved PCE plume, residual and/or free-phase PCE exists within the volume swept by the surfactant. The first is the demonstrated recovery of PCE at greater than the aqueous solubility for the five-drum ganglia case. This can only happen if the surfactant has solubilized PCE, or PCE has been recovered as a free-phase NAPL at the extraction well, or the surfactant solution has desorbed very large quantities of PCE from aquifer materials. The second point is the large increase in the recovered PCE concentrations and the spiked recovery corresponding to the recovery of the surfactant solution. Injecting surfactant solution into a dissolved PCE plume (i.e., no NAPL) would not cause increases in the PCE concentrations except by enhanced desorption of PCE from aquifer materials. Third, the PCE concentration during the initial sweep with ground water remains relatively constant, indicating a long-term source. If there were only a dissolved plume and sorbed PCE, the initial water flush would drop the PCE concentrations to very low values after one pore volume had passed through the system. Finally, the PCE concentrations in the water chaser remain at pre-surfactant solution injection values. If no pre-flush were performed, surfactant flushing of a zone with only a dissolved plume and sorbed PCE would create a relatively sharp peak of PCE recovery, followed by PCE concentrations near zero as both the plume and sorbed PCE will be flushed. Solubilization of residual or free-phase PCE, or extraction of free-phase PCE, is the only means for the order of magnitude increases in PCE recovery indicated while at the same time showing a long-term source.

Under most conditions, the amount of sorbed PCE will be a very small fraction of that contained in a NAPL zone. Consider a scenario with no PCE NAPL in the interwell region of the model, an organic carbon content of 0.1% in the aquifer material (f_{oc} = 0.001), a partitioning coefficient (K_{oc}) of 364 mL/g *(23)*, and an aqueous concentration of PCE at 10% of its aqueous solubility (10% = 24 mg/L). The volume of sorbed PCE contained in the interwell region would be approximately 2 liters plus approximately 1 liter of PCE dissolved in the ground water. Even with 100% PCE saturation in the ground water, the total volume of sorbed and dissolved PCE in the interwell region would only be approximately 30 liters. These PCE volumes are a small fraction of one drum and therefore cause interpretation problems only with small NAPL spills.

Figure 5 shows the cross section with the five drums of PCE distributed as a pool at the base of the aquifer. The screened interval extends from the top to the bottom of the cross section. Figure 6 shows a comparison of the PCE recovery at the extraction well from interwell simulations of the five-drum PCE pool and five-

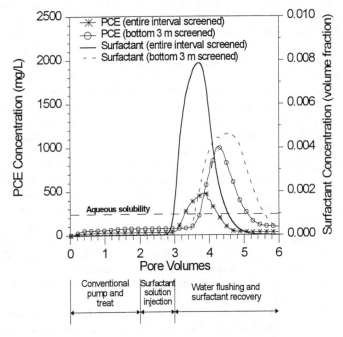

Figure 7. Comparison of results of interwell NAPL solubilization tests for the 5-drum PCE pool with the entire interval screened and with the bottom 3 meters screened.

drum PCE ganglia cases. Because the recovered concentrations of PCE are greater than the aqueous solubility, the results show that the interwell NAPL solubilization test does indicate residual or free-phase PCE in the test zone for both cases. In addition, significant dilution occurs due to the limited region for PCE solubilization in the case with the pool. However, even with the limited contact surface area for PCE solubilization of the pool, a clear signal of enhanced concentration for identifying the presence of the pool has been achieved.

Figure 7 shows the effect of changing the screened interval to decrease the dilution associated with the interwell test of the five-drum pool. By limiting the screened interval to the bottom 3 meters (9.8 ft) of the cross-section, the recovery of PCE is approximately twice the recovery when the entire 12 meters (39.4 ft) of the cross section is screened. In addition, each pore volume in the 3-meter screened interval case is one quarter of the pore volume for the 12-meter screened interval case. Therefore, one quarter of the surfactant amount was able to solubilize twice the PCE due to the reduced dilution. It can also be seen that the recovery for the 3-meter screened interval case is delayed due to the increased flow volume along the surfactant solution's flow path caused by the partially screened interval (allowing a vertical component in the flow path).

Summary and Conclusions

Numerical simulation has been used to demonstrate the NAPL solubilization test as a means to locate and characterize NAPL in the subsurface. The advantages of using a surfactant sweep in a well test are threefold. First, a significant volume of the aquifer can be tested for the presence of a NAPL when compared to the use of water samples and cores. Second, a positive indication of NAPL can be made due to recoveries of NAPLs above their aqueous solubilities or based on significant increases in NAPL recoveries associated with the surfactant sweep. Third, the chemical composition of the NAPL can be determined by appropriate choice of micellar surfactant solutions. The NAPL solubilization test can detect low-solubility hydrophobic NAPLs due to the surfactant's capability to increase their recovery. Knowledge of the NAPL's composition is critical in determining safety and treatment requirements for any future design of the appropriate remediation system.

Based on previous characterization techniques, much of the ongoing remediation work at NAPL-contaminated sites has been characterized by the location of the contaminant plume without regard to the location of the NAPL source. As a result, many remediation projects evolve into projects to contain the contaminant plume rather than projects to remove the NAPL, which Mackay and Cherry *(1, p. 632)* have pointed out is the "primary challenge in groundwater cleanup." By using single and interwell NAPL solubilization tests, the location and chemical composition of the NAPL source can be determined. This will permit remediation projects to focus on the NAPL zones, which is essential for the timely and cost-effective remediation of NAPL- contaminated sites.

Acknowledgments

We acknowledge the assistance of Chris Brown and Mojdeh Delshad of the University of Texas at Austin with the UTCHEM simulations. The University of Texas and INTERA acknowledge partial support for this work from EPA's R.S. Kerr Environmental Research Laboratory under Contract No. CR 821897-01-0.

Literature Cited

(1)   Mackay, D.M. and J.A. Cherry, **1989**.   Groundwater contamination: Pump-and-treat remediation. *Environ. Sci. Technol.*, 23(6):630-636.

(2)   Feenstra, S., D.M. Mackay, and J.A. Cherry, **1991**. A method for assessing residual NAPL based on organic chemical concentrations in soil samples. *Ground Water Monitoring Review*, Spring issue, pp. 128-136.

(3)   Cohen, R.M., A.P. Bryda, S.T. Shaw, and C.P. Spalding, **1992**. Evaluation of visual methods to detect NAPL in soil and water. *Ground Water Monitoring Review*, Fall issue, pp. 132-141.

(4)   Mayer, A.S. and C.T. Miller, **1992**. The influence of porous medium characteristics and measurement scale on pore-scale distributions of residual nonaqueous-phase liquids. *J. Contaminant Hydrol.*, 11: 189-213.

(5)   Jackson, R.E. and J.F. Pickens, **1994**.   Determining location and composition of liquid contaminants in geologic formations.   Patent No. 5,319,955, U.S. Patent Office, Washington, D.C.

(6)   EPA, **1992**. Considerations in Ground-Water Remediation at Superfund Sites and RCRA Facilities -- Update. Directive 9283.1-06, Office of Solid Waste and Emergency Response (OSWER), May 27, 1992.

(7)   EPA, **1992**. Dense Nonaqueous Phase Liquids -- A Workshop Summary. EPA/600/R-92/030, ORD, Washington, DC, 20460.

(8)   Reinhard, M., N. Goodman and J.F. Barker, **1984**.   Occurrence and distribution of organic chemicals in two landfill leachate plumes. *Environ. Sci. Technol.*, vol. 18, pp. 953-961.

(9)   McKinney, D.C., C. Tilburg, G.A. Pope, and K. Sepehrnoori, **1993**. Characterization of DNAPL zones: tracer tests and the inverse problem. *EOS, Transactions*, American Geophysical Union, vol. 74, no. 16, April 20, 1993/Supplement, p.124.

(10) Jin, M., M. Delshad, D.C. McKinney, G.A. Pope, K. Sepehrnoori, C. Tilburg, and R.E. Jackson, **1994**.   Subsurface NAPL contamination: partitioning tracer test for detection, estimation and remediation performance assessment. In: *Toxic Substances and the Hydrologic Sciences,* Ed. A.R. Dutton, American Institute of Hydrology, Minneapolis, MN, pp. 131-159.

(11) Fountain, J.C., **1992**. Field test of surfactant flooding: mobility control of dense nonaqueous-phase liquids.   In: *Transport and Remediation of Subsurface Contamination.* ACS Symposium Series 491, American Chemical Society, Washington, DC, chapter 15.

(12) Fountain, J.C., C. Waddell-Sheets, and R.E. Jackson, **1994**. Hydrogeologic considerations for remediation of NAPL-contaminated sites using surfactant-enhanced pump and treat. In: *Toxic Substances and the Hydrologic Sciences,* Ed. A.R. Dutton, American Institute of Hydrology, Minneapolis, MN, pp. 366-376.

(13) Kueper, B.H., D. Redman, R.C. Starr, S. Reitsma, and M. Mah, **1993**. A field experiment to study the behavior of tetrachloroethylene below the water table: spatial distribution of residual and pooled DNAPL. *Ground Water,* 31(5):756-766.

(14) Jackson, R.E., J.F. Pickens, A. Haug, J.C. Fountain, and S.H. Conrad, **1993**. Characterization and remediation of DNAPL zones in sand and gravel aquifers by chemically enhanced solubilization. In: *ER '93, Environmental Remediation Conference,* Augusta, GA, U.S. Department of Energy, Vol. 2, pp. 1261-1266.

(15) Bourrel, M. and R.S. Schechter, **1988**. *Microemulsions and related systems: formulation, solvency and physical properties.* Marcel Dekker, Inc., New York.

(16) Rosen, M.J., **1989**. *Surfactants and Interfacial Phenomena.* Second Edition. Wiley Interscience, New York.

(17) Edwards, D.A., R.G. Luthy, and Z. Liu, **1991**. Solubilization of polycyclic aromatic hydrocarbons by nonionic surfactant solutions. *Environ. Sci. and Technol.,* Vol. 25, pp. 127-133.

(18) Nagarajan, R. and E. Ruckenstein, **1984**. Selective solubilization in aqueous surfactant solutions. In: *Surfactants in Solution,* Vol. 2, edited by K.L. Mittal and B. Lindman, Plenum Press, New York, pp. 923-947.

(19) Butler, G.W., R.E. Jackson, J.F. Pickens, and G.A. Pope, **1994**. Surfactant-enhanced characterization of DNAPL zones. In: Preprints of Papers presented at the 207th ACS National Meeting, Division of Environmental Chemistry Division, American Chemical Society, Vol. 34, No. 1, pp. 628-631.

(20) Saad, N., **1989**. Field Scale Simulation of Chemical Flooding. Ph.D. dissertation, the University of Texas, Austin.

(21) Brown, C.L., **1993**. Simulation of Surfactant Enhanced Remediation of Aquifers Contaminated with Dense Non-Aqueous Phase Liquids. M.S. thesis, the University of Texas, Austin.

(22) Brown, C.L., G.A. Pope, L.M. Abriola, and K. Sepehrnoori, **1994**. Simulation of surfactant-enhanced aquifer remediation. *Water Resources Research,* Vol. 30, No. 11, pp. 2959-2978.

(23) Schwille, F., 1988. *Dense chlorinated solvents in porous and fractured media.* Translated by J.F. Pankow. Lewis Publishers.

RECEIVED December 15, 1994

Chapter 16

Surfactant-Induced Reductions of Saturated Hydraulic Conductivity and Unsaturated Diffusivity

B. Allred and G. O. Brown[1]

Department of Biosystems and Agricultural Engineering, Oklahoma State University, Stillwater, OK 74078

The loss in capability of a soil to transmit flow will decrease the efficiency of surfactant enhanced in situ environmental remediation. A soil's ability to transmit flow depends on its saturated hydraulic conductivity or unsaturated diffusivity. To investigate the effects on hydraulic conductivity, falling-head permeability tests were used. Four anionic and three nonionic surfactants were tested on two soils. The two different soil types utilized were a loam and a sand. Saturated hydraulic conductivity reductions due to the anionic surfactants ranged up to two orders of magnitude in the loam and 58% in the sand. With nonionic surfactants, maximum reductions were one order of magnitude in the loam and 44% in the sand. Transient unsaturated column tests were used to determine the effects on loamy soil diffusivity due to one of the anionic surfactants. In the unsaturated tests, maximum diffusivity reductions were up to one order of magnitude for volumetric moisture contents which were just below saturation.

Surfactant solution systems are presently being considered for use in flushing environmental contaminants from soils. The efficiency of using surfactants for this purpose may depend on how they affect the saturated hydraulic conductivity or unsaturated diffusivity. Large decreases in these soil properties may result in reducing efficiency to levels which make environmental remediation impractical. Evidence in the literature (1,2) suggests that surfactants can indeed cause significant conductivity/diffusivity reductions. This study attempts to quantify these reductions on two different soils using different surfactants.

---

[1]Corresponding author

0097–6156/95/0594–0216$12.00/0

Flow Theory

Darcy's law describing horizontal flow in saturated porous media can be written:

$$q = -K\frac{dh}{dx} \tag{1}$$

where q is specific discharge (L/T), K is saturated hydraulic conductivity (L/T), h is hydraulic head (L), and x is distance (L). In turn, saturated hydraulic conductivity can be expressed:

$$K = \frac{k\rho g}{\mu} \tag{2}$$

where the intrinsic permeability, k (L^2), is strictly a property of the porous media, ρ is fluid density (M/L^3), g is the gravitational acceleration constant (L/T^2), and μ is fluid viscosity (M/LT). From equation 2, it follows that surfactant solutions introduced into saturated porous media can alter conductivity values by changing fluid density and viscosity values. Surfactant-soil interactions which effect intrinsic permeability will also affect conductivity.

Horizontal flow in unsaturated porous media can be expressed:

$$\frac{\partial\theta}{\partial t} = \frac{\partial}{\partial x}\left(K(\theta)\frac{\partial h}{\partial x}\right) = \frac{\partial}{\partial x}\left(K(\theta)\frac{\partial\psi}{\partial x}\right) = \frac{\partial}{\partial x}\left(K(\theta)\frac{\partial\psi}{\partial\theta}\frac{\partial\theta}{\partial x}\right) = \frac{\partial}{\partial x}\left(D(\theta)\frac{\partial\theta}{\partial x}\right) \tag{3}$$

where t is time (T), ψ is the pressure head (L), and the unsaturated hydraulic conductivity, $K(\theta)$, is a function of the volumetric moisture content, θ. The unsaturated diffusivity, $D(\theta)$ (L^2/T), is defined as:

$$D(\theta) = \frac{k(\theta)\rho g}{\mu}\frac{\theta\psi}{\partial\theta} \tag{4}$$

Equation 4 implies that surfactant solutions introduced into unsaturated porous media can affect diffusivity values by altering the moisture content dependent intrinsic permeability, the pressure head versus moisture content relationship, or the fluid properties of density and viscosity.

Bruce and Klute (3) showed that equation 3 can be solved as an ordinary differential equation using a method devised by Boltzmann (4). A substitution of λ equal to $x/\sqrt{t}$ is used to transform equation 3 into the following form:

$$-\frac{\lambda}{2}\frac{d\theta}{d\lambda} = \frac{d}{d\lambda}\left(D(\theta)\frac{d\theta}{d\lambda}\right) \tag{5}$$

Table 1. Surfactant List

Chemical Name	Trade Name[1]	Abbreviation	Average Molec. Weight	Viscosity[2] (gm/cm-s) C=0.01 (mole/kg)	C=0.05 (mole/kg)
ANIONIC					
Na-Lauryl Sulfate	Witcolate A PWD	A1	288	0.0102	0.0114
Na-Alpha Olefin Sulfonate	Witconate AOS	A2	324	0.0101	0.0114
Na-Dodecyl Benzene Sulfonate	Aldrich Chemical Co. Cat. #28995-7	A3	348	0.0099	0.0111
Na-Laureth Sulfate (3EO)	Witcolate ES-3	A4	437	0.0101	0.0117
NONIONIC					
Alkyl Polyoxyethylene Glycol Ether	Witconol SN-90	N1	490	0.0103	0.0119
Alkylphenol Ethoxylate	Witconol NP-100	N2	640	0.0105	0.0152
Alkyl Polyoxyethylene Glycol Ether	Witconol 1206	N3	825	0.0105	0.0131

[1] All surfactants were obtained from the Witco Corp, with the exception A3.
[2] Temperature = 22 C. For water at T= 22 C, viscosity = 0.00956 gm/(cm-s).

Using laboratory tests in which a solution is injected into the inlet of a horizontally mounted soil column, the diffusivity versus moisture content relationship can be determined by rearranging equation 5 and integrating with respect to λ over the following boundary conditions:

$$\theta = \theta_i, \text{ for } \lambda \to \infty \ (x \to \infty \text{ or } t = 0)$$
$$\theta = \theta_0, \text{ for } \lambda = 0 \ (x = 0 \text{ and } t > 0) \tag{6}$$

where θ_i is the initial moisture content, and θ_0 is the inlet moisture content. The diffusivity relationship can then be defined as:

$$D\ (\theta_*) = -\frac{1}{2}\left(\frac{d\lambda}{d\theta}\right)_{\theta_*} \int_{\theta_i}^{\theta_*} \lambda\, d\theta \tag{7}$$

where θ_* is an arbitrary moisture content between θ_i and θ_0. The term, $d\lambda/d\theta$, represents the derivative at $\theta = \theta_*$. Given the testing conditions previously stated, equation 7 can be evaluated after determining the moisture content profile along the soil column.

Materials

Surfactants. Table I lists the surfactants used in this study along with their molecular weights and solution viscosities. The abbreviations provided in this table will be used to designate specific surfactants throughout the remainder of the text. Of the seven surfactants tested, four are anionic and three are nonionic. Anionic and nonionic surfactants are the types most likely utilized for environmental remediation. These particular surfactants were chosen because of their common commercial availability. Solution viscosities were obtained with a size 50 viscometer (Cannon Instrument Co.). Specific gravities of the tested surfactant solutions were essentially equal to 1.

Soil. The two soils tested in this study were chosen based on their initial saturated hydraulic conductivity values which are representative of extremes in suitability regarding in situ soil flushing remediation. The sand has an average initial permeability of 3×10^{-2} cm/s and would be an ideal candidate while the loam with average initial conductivity of 6×10^{-5} cm/s would be only marginally practical for soil remediation using flushing techniques. The Teller loam (Thermic Udic Argiustoll) and Dougherty sand (Thermic Arenic Haplustalf) were obtained from field locations near Perkins, Oklahoma. Characteristics of both soils are provided in Table II. Properties were determined using the procedures described in *Methods of Soil Analysis, Part 1 & 2* (*5,6*). Specific surface area was calculated from water vapor sorption isotherms using the B.E.T. formula (*7*). From Table II it is apparent that both soils are slightly acidic and calcium is the dominant exchangeable cation.

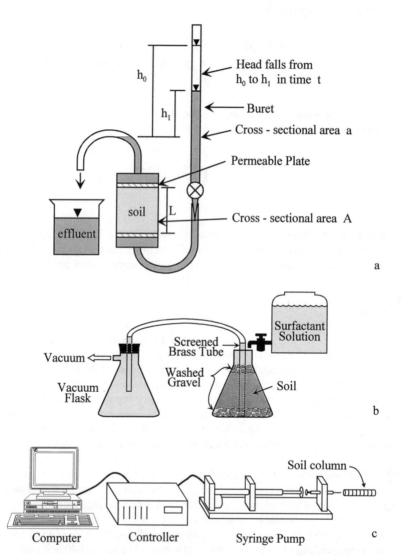

Figure 1. (a) Falling-head permeameter. (b) Dispersion test appartus.(c) Syringe pump equipment used for conducting unsaturated flow experments.

Testing Procedures

Saturated Tests. Changes in saturated hydraulic conductivity were monitored using standard falling-head permeability tests. The basic testing apparatus is shown in Figure 1a and the description of procedures is available in a number of introductory ground water hydrology texts (8-10). Saturated hydraulic conductivity was calculated using the following formula:

$$K = \frac{aL}{At} \ln \left(\frac{h_0}{h_1} \right) \tag{8}$$

where a is the cross-sectional area of the buret (L^2), A is the cross-sectional area of the soil column (L^2), L is the length of the soil column, and t is the time required for the hydraulic head to fall from h_0 to h_1. For these tests, the ratio of total head loss over column length (or hydraulic gradient) ranged from three to six. These values are consistent with common laboratory testing techniques but considerably higher than normal field conditions. All flow velocities are well within the range in which Darcy's law is valid, and matrix compression would be trivial at the packing densities used.

Rigid columns 4.15 cm in diameter and 15 cm in length were packed in uniform 1 cm lifts at dry bulk densities of 1.65 g/cm^3 for the Teller loam and 1.70 g/cm^3 for the Dougherty sand. These densities correspond to porosities of 38 and 36 percent, respectively. The soils were sieved and heated to 105 C for 24 hours prior to packing in order to inhibit microbial growth during testing. The effects of pore clogging microbial growth activated by surfactant biodegradation were considered beyond the scope of this study. The columns were then vacuum saturated with a deaerated, surfactant free, nominal soil water solution (0.001 mole/kg NaCl and 0.001 mole/kg CaSO$_4$). Because natural pore waters were not available from the soils tested, this solution, which contains low concentrations of common monovalent and divalent ions, was used to maintain an equilibrium in the soil chemistry.

Teller Loam Testing. After saturation, columns packed with Teller loam were flushed with approximately one pore volume of soil water solution. (A pore volume is equal to the total volume occupied by the voids within the soil column.) Initial saturated hydraulic conductivities were then obtained. Next, the surfactant solution was substituted into the buret and inlet supply tubing. After this substitution was accomplished, changes in conductivity were monitored with respect to the number of pore volumes of surfactant solution injected.

A series tests were conducted with the Teller loam to determine the effects of the seven individual surfactants along with two anionic-nonionic mixtures. Solution concentrations were 0.01 mole/kg (0.3 to 0.8 % by weight) which is the lower limit of what would be required for environmental remediation. Since solution systems used in soil flushing may require the benefits of having both anionic and nonionic surfactants present (11), two mixtures were tested in this study. The mixtures had a total combined concentration of 0.01 mole/kg and contained equal amounts of both the anionic and nonionic surfactants. For background comparison purposes, one test was run using water only.

Dougherty Sand Testing. Changes in the conductivity of the sand were measured with seven different surfactants and two mixtures. The anionic-nonionic mixtures contained equal amounts of both surfactants at a total combined concentration equal to that used for the separate individual surfactants tests (0.05 mole/kg). A control test with water was also conducted. Procedures for determining saturated hydraulic conductivity changes in the sand were somewhat different from those used for the loam. After saturation, the columns were flushed with approximately four pore volumes of soil water solution. Following the preflushing, initial saturated hydraulic conductivities were determined. Columns were then flushed with four pore volumes of 0.05 mole/kg surfactant solution and allowed to equilibrate for 12 hours. Next, a surfactant affected hydraulic conductivity was determined using the same solutions. The 0.05 mole/kg (1.5 to 4.0 % by weight) surfactant solution concentration approaches the upper limit of what is required for environmental remediation. The equilibration period was necessary to maintain a realistic duration of soil-surfactant contact and was chosen based on the calculated water travel time along a 15 cm (column length) flow path under a hypothetical field hydraulic gradient of 0.005.

During the Dougherty tests, some column effluent contained significant amounts of soil colloids which were not observed while testing the Teller. A simple procedure was devised to determine the relationship between effluent colloids and conductivity change. Figure 1b depicts the apparatus. A 300 cm^3 flask was filled with approximately 200 cm^3 of sand between two washed gravel layers. A brass tube open to the bottom gravel layer allowed effluent collection. The sand was preflushed by pouring four pore volumes of soil water solution into the top of the flask while applying suction to the brass tube. Then it was leached with an additional four pore volumes of 0.05 mole/kg surfactant solution and allowed to equilibrate for 12 hours. At the end of equilibration, one pore volume of effluent was extracted. This was followed by analysis of the effluent for nonvolatile solids (*12*), which is a measure of the soil colloids present.

Unsaturated Tests. The apparatus in Figure 1c was used to determine unsaturated diffusivity values. As shown, a computer controlled syringe pump was used to inject solution into the inlet of a dry soil column. Throughout the timed duration of the test, the volumetric moisture content at the inlet of the column was maintained at a constant value. The soil column itself was comprised of individual acrylic rings and packed with Teller loam to an average dry bulk density of 1.65 g/cm^3. The diameter of the column was 3.5 cm and the total length ranged from 10 cm to 30 cm. Upon completion of the test, the soil column was broken apart, and the moisture contents of the soil from within each ring were determined through oven drying at 105 C. After construction of a moisture content profile along the column, diffusivities were calculated using the mathematical techniques previously discussed. A more detailed account of testing procedures is provided by Brown and Allred (13).

Two types of testing were conducted. First, as a control, were a series of six tests, differing with respect to inlet moisture content, where water was injected into columns containing normal Teller loam soil. In a second series with seven tests, a 0.01 mole/kg A3 solution was injected into columns containing Teller loam soil which had been previously equilibrated with the A3 surfactant. Equilibration was accomplished by twice saturating and then draining a quantity of Teller loam soil using a 0.01 mole/kg solution of A3. The soil was then oven dried at 105 C prior to column packing. The purpose of equilibration was to saturate soil sorption sites with the surfactant. During testing, this would in turn maintain a 0.01 mole/kg A3 soil solution concentration along the moisture content profile from column inlet to wetting front edge. This procedure was necessary to insure that the diffusivities calculated for the second series of unsaturated tests were indeed surfactant affected values.

Results

Teller Loam Saturated Tests. Figure 2 depicts the hydraulic conductivity reductions in the loam caused by 0.01 mole/kg solutions of different surfactants and surfactant mixtures. Results of the four anionic surfactants are provided in Figure 2a. Of this group, surfactant A3 had the largest reduction which was close to two orders of magnitude over an injection of 0.3 pore volumes. Test data for the nonionic surfactants are shown in Figure 2b. Surfactant N2 exhibited a conductivity decrease of almost one order of magnitude over 1.5 pore volumes. This was the greatest reduction for the nonionic group. It is apparent from Figures 2a and 2b that anionic surfactants cause greater conductivity decreases than nonionics with respect to the loamy soil. Figure 2c illustrates the results of mixing anionic and nonionic surfactants. The mixed solution containing A1 and N3 behaved in a manner closest to the nonionic component, N3. The other mixed solution which contained surfactants A2 and N1, mimicked the behavior of the nonionic component, N1, for the first 1.4 pore volumes, but subsequently showed large conductivity reductions more like that of the anionic component, A2. In Figure 2, results of the test using water (0.001 mole/kg NaCl and 0.001 mole/kg $CaSO_4$) is provided for comparison purposes.

Dougherty Sand Saturated Tests. Table III presents the results of testing the Dougherty sand with 0.05 mole/kg solutions of seven individual surfactants and two mixtures. The largest decrease in conductivity was 58% for the A4 surfactant, while the smallest was 14 % for N1. On average, decreases were greater with anionics than nonionics. Anionic-nonionic mixtures showed conductivity decreases slightly more than those obtained for the straight nonionic solutions. Table III also presents the results of the soil colloid dispersion tests. Generally, the nonionic surfactants caused minimal dispersion, while anionic solutions and mixtures containing anionic surfactants produced effluent nonvolatile solids up to 7680 mg/kg. Clearly, anionic surfactant solutions have the greatest potential for dispersion and mobilization of colloids within the Dougherty sand.

Teller Loam Unsaturated Tests. Use of the previously discussed mathematical procedure for calculating unsaturated diffusivities is valid only if the moisture content profiles (θ versus $\lambda = x/\sqrt{t}$) show similarity for tests of different time duration but equivalent boundary conditions. Figure 3a shows this profile similarity in two sets of two tests. One of the sets was conducted with water and normal Teller loam while the other utilized 0.01 mole/kg A3 and surfactant equilibrated soil. Therefore, the unsaturated diffusivities calculated from both control and surfactant tests are valid. Also noteworthy from Figure 3a is that the leading edge of the moisture content profile

Table II. Soil Characteristics

Soil and USDA Classification	Extractable Bases (meq/100g)	Cation Exchange Capacity[1] (meq/100g)	pH	Specific Surface Area (m²/g)	Organic Carbon Content (weight%)
TELLER					
Loam	Na⁺ = 0.84	~14	6.0	37.8	1.2
52% Sand	K⁺ = 0.99				
31% Silt	Ca⁺² = 6.28				
17% Clay	Mg⁺² = 2.39				
DOUGHERTY					
Sand	Na⁺ = 1.40	~5	5.9	21.8	0.1
98% Sand	K⁺ = 0.14				
2% Sand and Clay	Ca⁺² = 2.40				
	Mg⁺² ~ 0.00				

[1] Cation Exchange Capacity was calculated assuming a base satuation of 75%, which is average for the Payne County, Oklahoma area.

Table III. Dougherty Sand Test Results

Injected Surfactant	Change in Hydraulic Conductivity %	Maximum Viscosity Effect Change %	Dispersed Nonvolatile Solids (mg/kg)
Water	5	0	176
A1	-47	-16	6750
A2	-35	-16	7680
A3	-54	-14	7439
A4	-58	-18	5471
N1	-14	-20	194
N2	-44	-37	125
N3	-22	-27	139
A1 & N3	-27	-25	2360
A2 & N1	-24	-17	5850

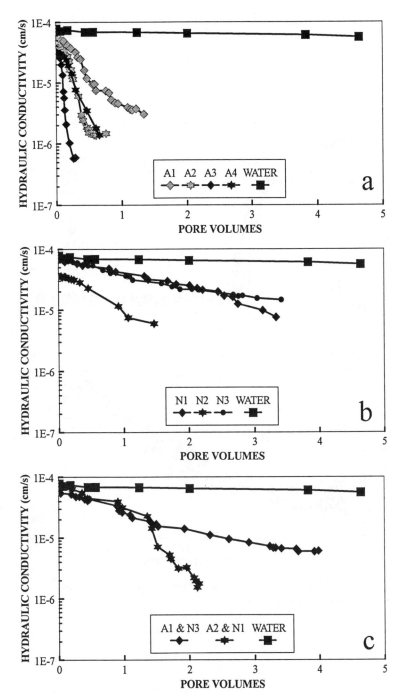

Figure 2. Teller loam saturated hydraulic conductivity versus the injected pore volumes of 0.01 mole/kg surfactant solutions. (a) Anionic surfactants. (b) Nonionic surfactants. (c) Anionic-nonionic surfactant mixtures.

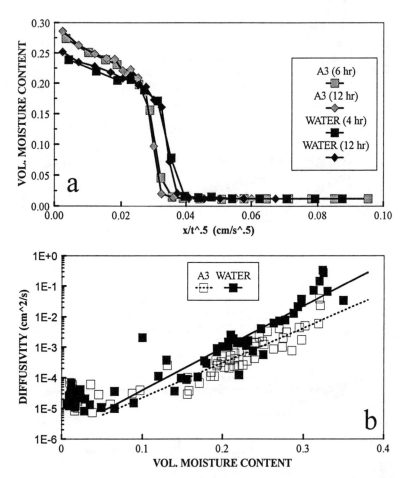

Figure 3. (a) Moisture content profiles plotted against $\lambda = x/\sqrt{t}$. (b) Diffusivity versus moisture content.

for the 12 hour surfactant test lagged significantly behind that of the 12 hour water test, although both tests had the same injection rates.

The moisture content profiles from six control and seven surfactant tests were used to calculate the diffusivities which are plotted in Figure 3b. Regression analysis was performed on both control and surfactant data points for a moisture content above 0.05. The depicted regression line for the control data is of the form:

$$D(\theta) = (10^{-5.8})e^{31.70} \tag{9}$$

Also shown is the surfactant data regression line which can be expressed:

$$D(\theta) = (10^{-5.8})e^{26.30} \tag{10}$$

The r^2 values calculated for the two equations given above are 0.82 and 0.83, respectively. At moisture contents less than 0.05, vapor transport dominates and therefore regression analysis was not extended below this level. The maximum calculated diffusivity reduction due to the A3 surfactant is almost one order of magnitude and occurs just below the saturation moisture content (0.38).

Discussion

Teller Loam Saturated Hydraulic Conductivity. By their nature, surfactants tend to be preferentially adsorbed at interfaces (*14*). Due to the large amount of interfacial area present in soils, the majority of the injected surfactant can probably be found adjacent to the column inlet. If so, this is the location where hydraulic conductivity reduction will occur. The measured surfactant affected conductivities are then likely to reflect effective values averaged over the whole column. The actual conductivity in the surfactant affected portion of the soil column will be less than this effective value.

The viscosities of the 0.01 moles/kg surfactant solutions listed in Table I are marginally greater than that of water. Equation 2 indicates that increasing the fluid viscosity will reduce hydraulic conductivity. However, hydraulic conductivity reductions in the Teller were measured in orders of magnitude losses and not the few percent which can be accounted for by differences in viscosity. Since changes in the fluid properties of density and viscosity are not responsible, equation 2 would indicate surfactant modification of intrinsic permeability to be the cause of conductivity reduction in the loam.

Anionic surfactants can precipitate as calcium salts in soils (*15*). The presence of calcium as the dominant cation in both tested soils makes the formation of pore clogging precipitates a strong possibility. This may be a significant mechanism by which anionic surfactants reduce conductivity, especially in the fine-grained loamy soil.

Precipitation would not be a dominant mechanism of conductivity reduction with nonionic surfactants. Here, the possibilities may include soil structure alteration caused by dispersion or even the formation of pore clogging lyotropic liquid crystals, which may form in response to a soil solution environment containing high levels of both inorganic electrolytes and residual organic matter. Mustafa and Letey (*16*) showed that two nonionic surfactants decreased soil aggregate stability. Miller et al.

(1) found that nonionic surfactants decreased flow rates in a hydrophobic soil and suggested that this phenomenon could be related to aggregate destabilization, micelle formation, or particle migration.

Dougherty Sand Saturated Hydraulic Conductivity. Differences in viscosity between the 0.05 mole/kg surfactant solutions and water (Table I) could account for much of the conductivity reductions observed in the sand. The reductions attributed to such viscosity effects are provided in Table III. Calculations indicate that viscosity effects may account for 90% of the reductions caused by nonionic surfactants and anionic-nonionic surfactant mixtures. For the anionic surfactants alone, this value is 33%. The remainder of conductivity reductions caused by anionics could result from mobilization of pore clogging soil colloids. The dispersion tests provide a good indication of the large amounts of soil colloids capable of being mobilized by anionic surfactants.

It is entirely possible, however, that the viscosity effects discussed above are in fact minimal. This would be the case if surfactants are removed from the soil solution through strong adsorption onto solid particles. The bulk soil solution would then have a viscosity similar to that of water. If so, other factors would be responsible for the conductivity reductions in the sand.

Teller Loam Unsaturated Diffusivity. Equation 4 indicates that diffusivity is dependent on the porous media intrinsic permeability, the fluid properties of density and viscosity, and the gradient between pore pressure and moisture content. For the 0.01 mole/kg A3 injection solution, the fluid properties of density and viscosity would not be a factor in reducing unsaturated diffusivity values. Diffusivity reductions could be the result of A3 surfactant effects on intrinsic permeability. Intrinsic permeability decreases could be caused by a number of factors which include surfactant precipitation as a calcic salt and soil structure alteration due to surfactant induced aggregate destabilization.

Because of surfactant reduced surface tensions, the pressure head versus moisture content gradient may decrease. This assumes that the cosine of the contact angle between the soil solution and the solid porous media does not increase more than surface tension decreases when surfactants are added. Therefore, unsaturated diffusivity reductions may be due to decreases in both intrinsic permeability and the pressure head versus moisture content gradient. At present it is unclear which factor dominates.

Conclusions

The saturated hydraulic conductivity reductions observed in this study vary significantly with respect to the different surfactants and the two soils tested. Because of its low initial conductivity, the Teller loam would be a poor candidate for soil flushing remediation. For this soil, the hydraulic conductivity reductions of almost two orders of magnitude caused by the 0.01 mole/kg A3 solution would make most surfactant flushing remediation impractical.

Hydraulic conductivity losses in the sand were also significant but not nearly as great as in the loam. Viscosity effects, if not negated by sorption processes, could account for much of the decrease. The Dougherty sand is representative of an ideal candidate for surfactant flushing remediation based on initial conductivity. The measured hydraulic conductivity reductions in the sand would decrease efficiency but not to the extent of making surfactant enhanced remediation unrealistic. In terms of minimizing the amount of saturated hydraulic conductivity reduction in both types of soil, nonionic surfactants appear to be a far better choice in comparison with anionic surfactants.

Surfactants can also affect unsaturated diffusivity values. A one order of magnitude decrease in the Teller loam was caused by the A3 surfactant at a moisture content approaching saturation. These diffusivity reductions probably reflect decreases in both intrinsic permeability and the pressure head versus moisture content gradient. This surfactant effect will need to be taken into account before initiating in situ flushing of unsaturated soils.

Acknowledgments

Support for this research was provided by the Oklahoma Agricultural Experiment Station as a contribution to Project 2151, and this paper is published with the approval of its director.

Authors Note

The use of brand names in this paper is for identification purposes only and does not constitute endorsement by the authors or Oklahoma State University.

Literature Cited

(1)   Miller, W. W.; Valoras, N.; Letey, J. *Soil Sci. Soc. of Am. Proc.* **1975**, *39*, 12-16.
(2)   Nash, J.; Traver, R. P.; Downey, D. C. *Surfactant Enhanced In Situ Soil Washing*; U. S. Air force Engineering and Services Center: Tyndal Air Force Base, FL, 1987; Report no. ESLTR-87-18.
(3)   Bruce, R. R.; Klute, A. *Soil Sci. Soc. of Am. Proc.* **1956**, *20*, 458-462.
(4)   Boltzmann, L. *Ann. Physics.* **1894**, *53*, 959-964.
(5)   *Methods of Analysis, Part 1 - Physical and Mineralogical Properties, 2nd ed.*; ASA and SSSA: Madison, Wisconsin, 1986.
(6)   *Methods of Analysis, Part 2 - Chemical and Microbiological Properties, 2nd ed.*; ASA and SSSA: Madison, Wisconsin, 1982.
(7)   Quirk, J. P. *Soil Science.* **1955**, *80*, 423-430.
(8)   Todd, D. K. *Groundwater Hydrology 2nd ed.*; John Wiley & Sons: New York, NY, 1980.
(9)   Freeze, R. A.; Cherry, J. A. *Groundwater*; Prentice-Hall, Inc.: Englewood Cliffs, NJ, 1979.
(10)   McWhorter, D. B.; Sunada, D. K. *Ground-Water Hydrology*; Water Resources Publications: Lakewood, CO, 1977.
(11)   American Petroleum Institute. *Test Results of Surfactant Enhanced Gasoline Recovery in a Large-Scale Model Aquifer*; 1985: API Publication No. 4390.

(12)   *Standard Methods for the Examination of Water and Wastewater, 16th ed.*;
       APHA, WWA, and WPCF: Washington, D. C., 1985.
(13)   Brown, G. O.; Allred, B. *Soil Science.* **1992**, *154*, 243-249.
(14)   Rosen, M. J. *Surfactants and Interfacial Phenomena, 2nd ed.*; Wiley
       Interscience:New York, NY, 1989.
(15)   Decreux, J.; Bocard, C.; Muntzer, P.; Razakarisoa, O.; Zilliox, L. *Water Sci.
       Tech.* **1990**, *22*, 27-36.
(16)   Mustafa, M. A.; Letey, J. *Soil Science.* **1969**, *107*, 343-347.

RECEIVED December 13, 1994

Chapter 17

Recovery of a Dialkyl Diphenyl Ether Disulfonate Surfactant from Surfactant Flush Solutions by Precipitation

Yuefeng Yin, John F. Scamehorn[1], and Sherril D. Christian

Institute for Applied Surfactant Research, University of Oklahoma, Norman, OK 73019

Surfactant precipitation is one method of separating and concentrating surfactant for reuse from a subsurface surfactant-based remediation process. One class of surfactants which has been shown to be very effective in this application is dialkyldiphenylether disulfonates. In this work, the precipitation and coacervation phase boundaries for a surfactant mixture which is primarily didecyldiphenylether disulfonate as a function of concentration of added NaCl and KCl are reported. The Krafft temperature (lowest temperature at which precipitation occurs) is only mildly affected by the added electrolyte. The fraction of surfactant precipitated is shown to be relatively small except at very high added electrolyte levels. The rate of precipitation is slow except at very high added electrolyte levels; e.g., more than a week is sometimes required for equilibration. One of the most attractive aspects of these surfactants for use in subsurface remediation is their low tendency to precipitate. This is a disadvantage when precipitation is necessary for surfactant recovery.

In order for surfactant-enhanced subsurface remediation operations to be an economic solution to soil contamination, recovery of the surfactant for reuse is necessary (1). Prior to separating the organic or ionic pollutant from the surfactant, these components can be concentrated in the

---

[1]Corresponding author

0097–6156/95/0594–0231$12.00/0
© 1995 American Chemical Society

flush solution using micellar-enhanced ultrafiltration (MEUF) (2-4). In MEUF, the solution is treated by ultrafiltration with membrane pore sizes small enough to block the passage of micelles with solubilized organic pollutants or bound ionic contaminants (of opposite charge to that of the surfactants). The permeate solution passing through the membrane can be reinjected into the aquifer since it contains low surfactant concentrations. The retentate solution not passing through the membrane can be treated to separate the surfactant from the pollutant to permit reuse of the surfactant. A typical retentate solution contains about 0.3 M surfactant, since above that concentration, concentration polarization causes flux through the membrane to be unacceptably low. For nonionic surfactants, an alternative to MEUF is to heat the solution above the cloud point where the solution phase separates into a dense coacervate solution containing a high concentration of surfactant and pollutant and a dilute solution which can be reinjected (5). There are several options for this retentate (or coacervate) treatment step which are highly dependent on the nature of the surfactant and the pollutant.

(A): An ionic surfactant can be precipitated by addition of electrolyte or reduction of temperature. This surfactant can then be separated by filtration, centrifugation, or gravity settling. The surfactant can then be redissolved for reuse. The pollutant may phase separate into a separate liquid phase which can be skimmed off as its solubility decreases as the surfactant is removed from solution upon precipitation. If the pollutant forms a dense liquid phase or a dense precipitate, this technique is not effective. This method is discussed for a specific surfactant in this paper and has been analyzed for other surfactants in other publications (6,7).

(B): If the pollutant is a volatile organic solute, the retentate can be vacuum, air, or steam stripped (4,8,9). In this case, the pollutant is removed overhead from the stripper and is condensed for disposal as a liquid or fed directly to an incineration unit. The bottoms stream from the stripper contains the surfactant in concentrated form, which can be directly reused.

(C): If the pollutant is a metal or metallic complex (e.g., zinc or chromate), electrolyte can be added to solution or the pH adjusted to cause the pollutant to be precipitated from solution. After removal of the solid precipitate for disposal, the remaining surfactant solution can be reused.

(D): An organic extractant phase can be contacted with the retentate to extract the pollutant or alternatively the surfactant. The solvent requires regeneration (e.g., distillation) for reuse. Back extraction of the solvent into the aqueous solution is also of concern, particularly if surfactant is left in this aqueous solution to allow micellar solubilization of the solvent.

Surfactant precipitation would be the most generally applicable of these methods since it is not as affected by the nature of the pollutant and mixed organic/ionic pollutants could be treated. It is particularly desirable to produce large, dense precipitate crystals which settle out of solution readily so that inexpensive gravity settlers can be used to remove the crystals from the solution. Use of divalent counterions (e.g., calcium for anionic surfactants) tends to lead to colloidal precipitate (6,10), so this work has emphasized monovalent counterions.

A new class of surfactants referred to as "Gemini" surfactants (11-14) have the characteristics of a hydrocarbon chain, an ionic group, a spacer, a second ionic group, and another hydrocarbon tail. The linear didecyldiphenylether disulfonate (C10-DADS) used here involve an ether oxygen as a flexible spacer. The dialkyldiphenylether disulfonates have been shown to exhibit excellent properties for subsurface remediation, including a low tendency to precipitate and a low adsorption onto soils (15). These surfactants have also been shown to be very promising for use in groundwater remediation using MEUF (4). Therefore, this work investigates the use of precipitation to recover a typical surfactant from this class from aqueous solution.

Experimental

Materials. The surfactant used in this work was a research sample supplied by Dow Chemical Company (XU40490.75). This sample was concentrated in C10-DADS, the primary component in the commercial product DOWFAX 3B2. The sample was reported to contain 28.5 wt. % C10-DADS, 2 wt. % monoalkyldiphenylether disulfonate, and 1 % other surfactants (other isomers, homologues, monosulfonates, etc.) which were unspecified, 1.1 wt. % Na_2SO_4, and water making up the remainder. The alkylation was reported as >98% pure and 96.5 % of the sulfonation product were disulfonates. The samples were desalted (99.3 % removal of NaCl and Na_2SO_4) using ultrafiltration with a membrane of molecular weight cut-off of 500 Daltons (small enough to block the surfactant monomer as well as micelles) at 40°C. Sodium chloride and potassium chloride were reagent grade. Since these materials were hygroscopic, both crystals were dried in a oven at 240°C for six hours for dehydration. Water used in all the experiments was double deionized and treated by a carbon filter.

Methods. To determine the Krafft temperature, surfactant sample solutions at concentrations above the CMC were first subcooled long enough to allow precipitation to occur. The

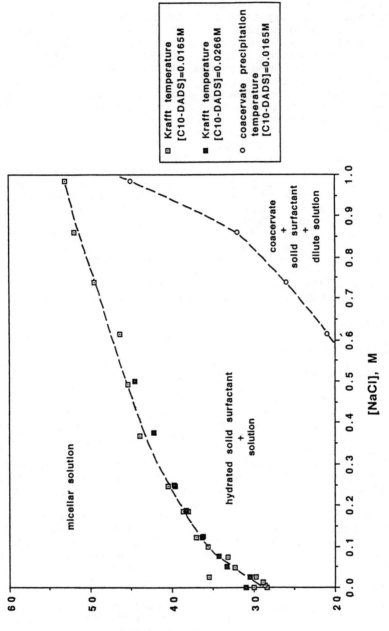

Figure 1. Effect of added NaCl on phase boundaries of C10-DADS solutions.

temperature was then raised incrementally with the interval about 2°C when T<10°C, about 1.5°C when 10°C<T<15°C, about 1°C when 15°C<T<25°C and about 0.3°C when T>25°C. The samples were kept at least 60 minutes at each temperature to reach equilibrium. The samples were observed visually with the aid of a flash light and were shaken about a hundred times for each of the several measurements at each temperature. The Krafft points of the samples were the temperature at which the crystals quickly dissolved (16-20). The concentration of unprecipitated C10-DADS was measured by UV spectroscopy at a wavelength of 232 nm.

In the study of precipitation kinetics, all the samples were subcooled at 0°C and allowed to settle. The samples were observed intermittently and the amount of precipitate was recorded as height and relative volume (volume of precipitate/volume of solution, which is equal to the height of precipitate/height of solution for the uniform diameter test tubes used here). Samples were observed for twenty-one days.

Results and Discussion

Phase Boundaries and Krafft Temperatures. The equilibrium phase boundaries of C10-DADS/NaCl solutions at two concentrations, both well above the CMC (CMC values are discussed later in this section), are shown in Figure 1. The Krafft temperature represents the highest temperature at which precipitated surfactant (hydrated solid surfactant) is present. The Krafft temperature is not significantly affected by surfactant concentration, as expected if the concentration is well above the CMC (21), at least for single component systems. The Krafft temperature increases slowly with added NaCl concentration, increasing by only about 25°C when 1 M NaCl is present relative to no added salt. At higher salinities, coacervate (a viscous surfactant-rich phase) is observed simultaneously with precipitate.

The equilibrium phase boundaries of C10-DADS/KCl solutions at two concentrations, both well above the CMC, are shown in Figure 2. The Krafft temperatures from Figure 2 are replotted in Figure 3 as a function of the mole fraction of potassium composing the counterions (e.g., $[K^+]/([K^+]+[Na^+])$). At low added KCl concentrations, the sodium salt of the surfactant is the precipitating species. The added KCl depresses the CMC of the surfactant by reducing the electrostatic repulsion between head groups at the micelle surface. This reduces the C10-DADS monomer concentration, lowering the Krafft temperature or temperature at which the solubility product of the Na/C10-DADS is exceeded. At higher KCl concentrations, the potassium salt of the surfactant is the precipitating species and the Krafft temperature increases with further increases in

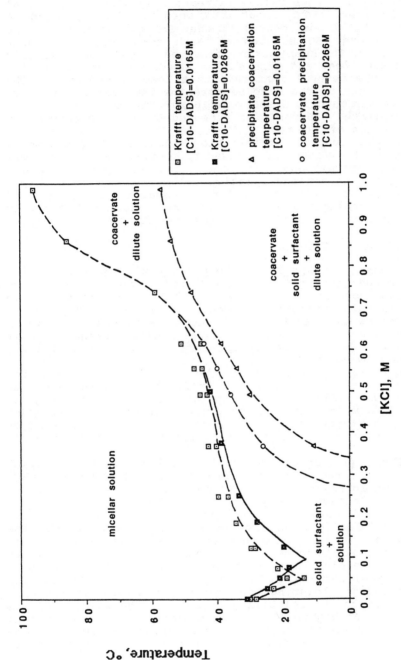

Figure 2. Effect of added KCl on phase boundaries of C10-DADS solutions.

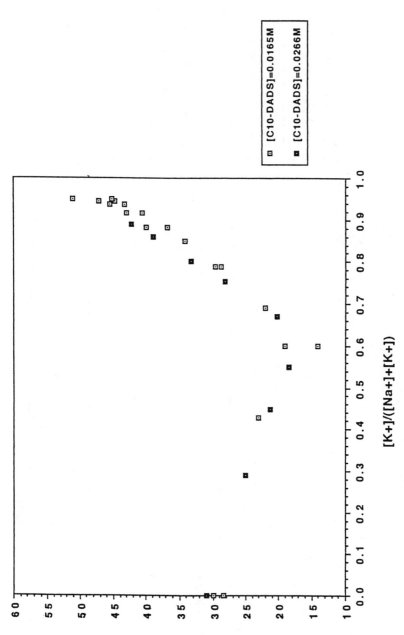

Figure 3. *Krafft temperature of C10-DADS solutions as a function of counterion composition.*

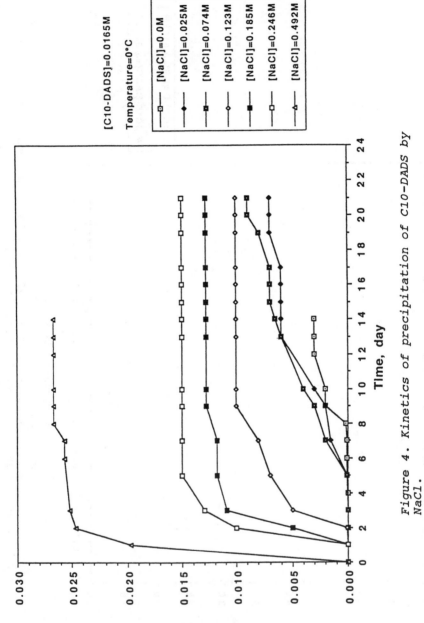

Figure 4. Kinetics of precipitation of C10-DADS by NaCl.

[KCl]. This results in a eutectic type curve as seen in Figure 3 or the minimum in the Krafft temperature in Figure 2. This eutectic behavior has been observed in other binary anionic surfactant systems (21-24). In this system, there is a region in the phase diagram where only coacervate and dilute solution (no precipitate) is present. At even higher KCl concentrations, precipitate and coacervate can be simultaneously present. As with NaCl as the additive, fairly high concentrations of KCl need to be added to the C10-DADS to substantially affect the Krafft temperature, although the dependence of Krafft temperature on [KCl] becomes greater than for NaCl above about 0.6 M added salt.

Since it is important to know the CMC at different salinities to be able to interpret and understand phase diagrams, the Corrin-Harkins equation (25) was used to correlate CMC data in this work:

$$\log_{10}(CMC) = -0.4269 \ \log_{10}[Na+] - 4.275 \qquad (1)$$

$$\log_{10}(CMC) = -0.5329 \ \log_{10}[K+] - 4.369 \qquad (2)$$

where the CMC, [Na+], and [K+] are in M. Equation 1 applies at 36°C and Equation 2 applies at 25°C. Precipitation disallowed both systems being at 25°C. The measured surface tension data and CMC values derived from those data are reported in reference (26).

Kinetics of Precipitation and Coacervation. The rate of precipitation and coacervation for the C10-DADS/NaCl system is shown in Figures 4 and 5 and for the C10-DADS/KCl system in Figures 6 and 7. From Figure 4, the precipitation takes about a day to reach equilibrium for high (ca. 0.5 M) NaCl concentrations, but can take weeks for lower concentrations, all at 0°C. Coacervation is fast for this system, reaching equilibrium in less than a day. These results are not inconsistent because coacervation only occurs at high NaCl concentrations (see Figure 1).

From Figure 6, the rate of precipitation with added KCl is slow, even at high added salt levels requiring more than a week to reach equilibrium. At low levels of added salt, equilibrium was not reached after several weeks. However, the rate of coacervation is fairly rapid as seen in Figure 7 for this C10-DADS/KCl system.

Fraction of Surfactant Precipitated. The solution concentration of the surfactant in equilibrium with precipitate is shown at several temperatures and salinities for NaCl in Figure 8. The surfactant solution concentration in equilibrium with precipitate or coacervate for KCl systems is shown in Figure 9. Comparing Figures 8 and 9 with Figures 1 and 2, the solutions are below the Krafft temperature (solution in

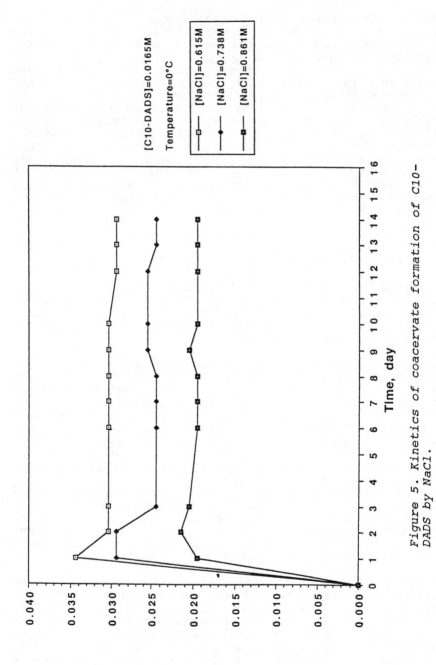

Figure 5. Kinetics of coacervate formation of C10-DADS by NaCl.

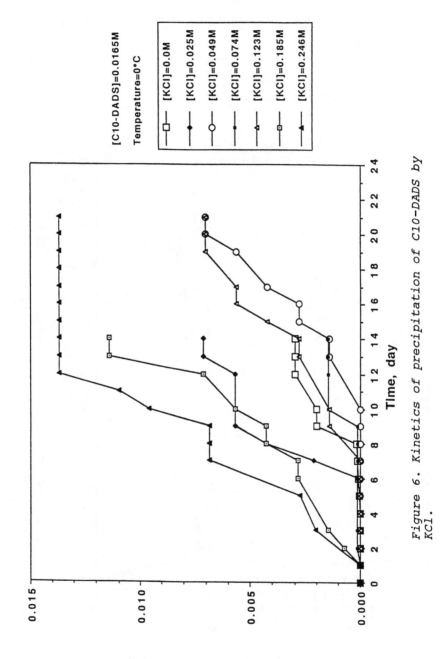

Figure 6. *Kinetics of precipitation of C10-DADS by KCl.*

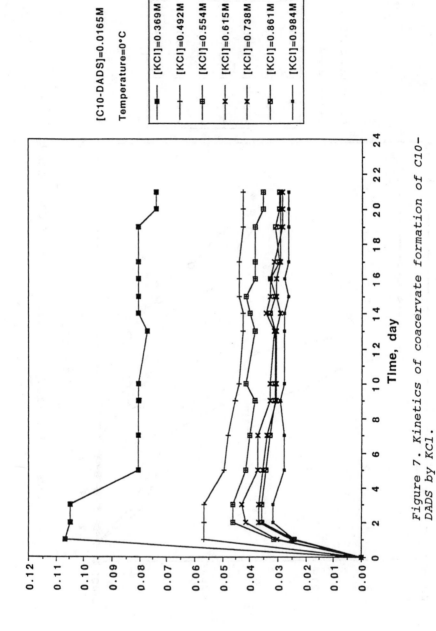

Figure 7. Kinetics of coacervate formation of C10-DADS by KCl.

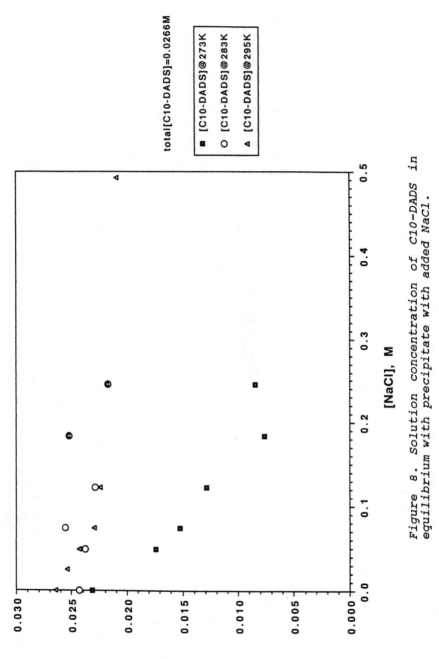

Figure 8. Solution concentration of C10-DADS in equilibrium with precipitate with added NaCl.

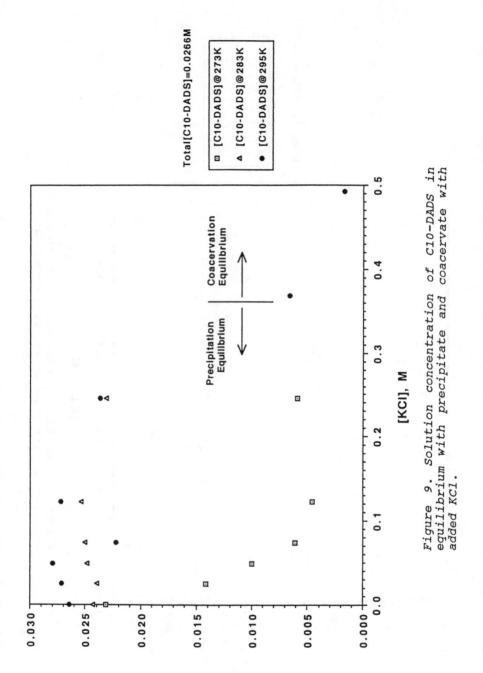

Figure 9. Solution concentration of C10-DADS in equilibrium with precipitate and coacervate with added KCl.

equilibrium with precipitate) under all conditions in Figure 8. For the added KCl system in Figure 9, either precipitate or coacervate is present at all conditions except for a narrow region between about 0.05 and 0.15 M added KCl for 295°K, where only a micellar solution is present.

The solubility of a single surfactant system decreases dramatically at temperatures below the Krafft temperature to the CMC at that temperature (27). The lowest solution concentrations shown in Figures 8 and 9 are well above the CMC values from Equations 1 and 2, in most cases well over an order of magnitude greater. This discrepancy is probably due to the highly heterogeneous nature of the surfactant mixture used. At the Krafft temperature, only the most easily precipitated surfactant is salted out of solution (21). It may take substantially lower temperatures for some of the other surfactant components to precipitate. The lowest temperature and highest salinities noted in Figures 8 and 9 for solution-precipitate equilibrium corresponds to about 80 % of the surfactant present being precipitated or 80 % maximum surfactant recovery from the stream. Somewhat lower surfactant concentrations remain in solution after coacervation, but formation of coacervate is not an effective way to separate surfactant from pollutant, since both components tend to concentrate in the coacervate (5).

Effect of Organic Pollutant. The effect of organic pollutants on phase behavior of anionic surfactant systems is expected to be highly dependent on the concentration and structure of the pollutant. In this work, the effect of 0.0025 M o-chlorophenol on the phase diagrams and on the kinetics of precipitation was measured (see Reference 26 for details). The presence of the pollutant only reduced the Krafft temperature by a maximum of 2°C. The presence of a pollutant could either increase or decrease the rate of precipitation, depending on the concentration of added electrolyte (either NaCl or KCl).

Implications for Process Design. Precipitation of anionic surfactant for reuse from a stream (either a subsurface flush solution or retentate from a MEUF treatment of such a stream) can be done by either reducing temperature or by adding electrolyte. The latter option would be preferable due to the cost of cooling a large stream. The optimum surfactant system would have a Krafft temperature which is very dependent on salinity, a rapid rate of precipitation, and dense crystals of surfactant which would settle by gravity. This system should also only contain a single isomerically pure surfactant and a low CMC so that the unprecipitated surfactant concentration is low after the crystallization, or fraction of surfactant recovered is high.

In the envision process, a small concentration of electrolyte would be added to the stream, causing the surfactant to precipitate and settle out. After separation by settling, the surfactant would be redissolved by dilution in water for reuse. The excess liquid after the settling would be treated to remove the organic and/or ionic pollutant, the process used being highly dependent on the nature of the pollutant (e.g., carbon adsorption, vacuum stripping, or phase separation of insoluble organics). The residual water remaining after pollutant removal and the recovered surfactant solution could be reinjected into the aquifer being remediated.

The C10-DADS used here is intermediate between an isomerically pure system and a commercial product, which would be even more heterogeneous. A purer surfactant would have a much higher percentage surfactant precipitated than observed here. It is also expected that a purer sample would precipitate faster than observed here, although this is somewhat speculative since very few studies have investigated kinetics of surfactant precipitation (21,28-30). However, a purer sample would probably still have a mild dependence of Krafft temperature on salinity as observed in Figures 1 and 2. It is this resistance to precipitation, among other properties, which makes this class of surfactants so desirable in subsurface remediation and makes them difficult to recover by precipitation above ground.

The slow rate of precipitation and low fraction of surfactant precipitated below the Krafft temperature make the surfactant sample studied here a poor candidate for recovery by precipitation. A monoisomeric sample of the same surfactant type may have much better properties for recovery. The degree of improvement associated with decreasing heterogeneity and whether this would offset the anticipated increased manufacturing cost are questions which require further research to resolve.

Acknowledgements

Financial support for this work was provided by the Office of Basic Energy Sciences, Department of Energy, Contract DE-FG05-87ER13678, Department of Energy Grant No. DE-FG01-8-7FE61146, National Science Foundation Grant CBT 8814147, an Applied Research Grant from the Oklahoma Center for the Advancement of Science and Technology, the Center for Waste Reduction Technologies of the American Institute of Chemical Engineers, Agreement No. N12-N10, and the TAPPI Foundation. In addition, support was received from the industrial sponsors of the Institute for Applied Surfactant Research including Amway, Amoco, Aqualon, Calgon-Vestal, Dow, International Paper, Kerr-McGee, Sandoz, DuPont, Unilever, Union Carbide, ICI, Shell, and Phillips Petroleum. The surfactant used was provided by Dow Chemical Co.

Literature Cited

1. Krebbs-Yuill, B.; Harwell, J.H.; Sabatini, D.A.; Knox, R.C. ACS Symp. Ser.; In Press - see as a chapter in this volume.
2. Christian, S.D.; Scamehorn, J.F. In "Surfactant-Based Separation Processes"; Scamehorn, J.F.; Harwell, J.H., Eds.; Marcel Dekker: New York, 1989; Ch. 1.
3. Scamehorn, J.F.; Christian, S.D. In "Surfactant-Based Separation Processes"; Scamehorn, J.F.; Harwell, J.H., Eds.; Marcel Dekker: New York, 1989; Ch. 2.
4. Roberts, B.L. PhD Dissertation, University of Oklahoma, Oklahoma, 1993.
5. Gullickson, N.D.; Scamehorn, J.F.; Harwell, J.H. In "Surfactant-Based Separation Processes"; Scamehorn, J.F.; Harwell, J.H., Eds.; Marcel Dekker: New York, 1989; Ch. 6.
6. Brant, L.W.; Stellner, K.L.; Scamehorn, J.F. In "Surfactant-Based Separation Processes"; Scamehorn, J.F.; Harwell, J.H., Eds.; Marcel Dekker: New York, 1989; Ch. 12.
7. Wu, B. M.S. Thesis, University of Oklahoma, Oklahoma, 1994.
8. Choori, U. M.S. Thesis, University of Oklahoma, Oklahoma, 1994.
9. Hasagawa, M.S. Thesis, University of Oklahoma, Oklahoma, 1994.
10. Peacock, J.M.; Matijevic, E. J. Colloid Interface Sci. 1980, 77, 548.
11. Menger, F.M; Littau, C.A. J. Am. Chem. Soc. 1991, 113, 1451.
12. Menger, F.M.; Littau, C.A. J. Am. Chem. Soc. 1993, 115, 10083.
13. Rosen, M.J. Chemtech 1993, 30.
14. Rosen, M.J.; J. Am. Oil Chem. Soc. 1992, 69, 30.
15. Shiau, B.J.; Rouse, J.D.; Soerens, T.S.; Sabatini, D.A.; Harwell, J.H. ACS Symp. Ser.; In Press - see as a chapter in this volume.
16. Stellner, K.L.; Scamehorn, J.F. J. Am. Oil Chem. Soc. 1986, 63, 566.
17. Stellner, K.L.; Scamehorn, J.F., Langmuir 1989, 5, 70.
18. Stellner, K.L.; Scamehorn, J.F., Langmuir 1989, 5, 77.
19. Stellner, K.L.; Amante, J.C.; Scamehorn, J.F.; Harwell, J.H. J. Colloid Interface Sci. 1988, 123, 186.
20. Amante, J.C.; Scamehorn, J.F.; Harwell, J.H. J. Colloid Interface Sci. 1991, 144, 243.
21. Scamehorn, J.F.; Harwell, J.H. In "Mixed Surfactant Systems"; Ogino, K.; Abe, M., Eds.; Marcel Dekker: New York, 1993; Ch. 10.
22. Scamehorn, J.F. ACS Symp. Ser. 1992, 501, 392.
23. Tsujii, K.; Saito, N.; Takeuchi,T. J. Phys. Chem. 1980, 84, 2287.

24. Hato, M.; Shinoda, K. J. Phys. Chem. 1973, 77, 378.
25. Corrin, M.L.; Harkins, W.D. J. Am. Chem. Soc. 1947, 69, 684.
26. Yin, Yuefeng M.S. thesis, University of Oklahoma, Oklahoma, 1994.
27. Shinoda, K. In "Colloidal Surfactants"; Shinoda, K.; Tamamushi, T.; Nakagawa, T.; Isemura, T., Eds.; Academic Press: New York, 1963, Ch. 1.
28. Clarke, D.E.; Lee, R.S.; Robb, I.D. Faraday Discus. Chem. Soc. 1976, 61, 165.
29. Lee, R.S.; Robb, I.D. Faraday Discus. Chem. Soc. 1979, 75, 2116.
30. Lee, R.S.; Robb, I.D. Faraday Discus. Chem. Soc. 1979, 75, 2126.

RECEIVED December 15, 1994

Chapter 18

Modeling the Effectiveness of Innovative Measures for Improving the Hydraulic Efficiency of Surfactant Injection and Recovery Systems

Y. Chen[1], L. Y. Chen[1], and R. C. Knox[1,2]

[1]School of Civil Engineering and Environmental Science and
[2]Institute for Applied Surfactant Research, University of Oklahoma, Norman, OK 73019

Vertical circulation wells (VCWs) can improve the hydraulic efficiency of surfactant injection/extraction systems used for remediation of subsurface contamination by dense non-aqueous phase liquids (DNAPLs). Simulations using a two-dimensional, glass plate sand tank allowed for visual observation of the flow mechanics of the VCW system and development of gross quantitative measures regarding performance. The VCW system was tested using both surfactant-based remediation mechanisms - solubilization and mobilization. The same tests were completed using a traditional two well (injection/extraction) system. The mobilization mechanism can remove more DNAPL mass per volume of surfactant than solubilization; however, to remain more effective, the mobilized DNAPL mass must be removed before it reaches any diffusion limited zones. The VCW system performs better than the two-well system using either mechanism. The VCW system also provides for more complete recovery of the injected surfactant solution.

Concern about subsurface contamination by dense non-aqueous phase liquids (DNAPLs) is widespread because of their existence at a large number of sites, their persistence in the subsurface as trapped residual and/or separate phases, and their ability to contaminate a very large volume of ground water. DNAPLs are denser than water (specific gravity > 1), generally of low viscosity, and are only sparingly soluble in water (*1*). The Maximum Contaminant Levels (MCLs) for most DNAPLs are at least two orders of magnitude less than their aqueous solubility. The maximum volume of ground water that ultimately may be contaminated by any given spill increases as the ratio of solubility to the volume of contaminant released decreases (*2*). Some DNAPLs can be of higher viscosity than water and can have high water solubility (*1*). Due to such conflicting characteristics, DNAPL contaminated subsurface environments are extremely difficult to remediate.

0097–6156/95/0594–0249$12.00/0
© 1995 American Chemical Society

Efforts at remediating subsurface organic chemical (especially DNAPL) contamination have been characterized as being costly, time-consuming and ineffective (3). The inefficiency of conventional pump-and-treat methods for remediating residual saturation or highly hydrophobic organics has been addressed by several recent reviews. Keeley (4) lists desorption of contaminants from media surfaces and liquid partitioning of immiscible contaminants as limiting factors. Haley et al. (5) determined that containment, rather than remediation, of organic contaminants was usually achieved using pump-and-treat methods. High sorption of organics and residual saturation were again cited as limiting factors. The authors recommended that research focus on methods to enhance extraction of these contaminants from the subsurface.

Surfactants are one class of chemical agents that can alter the physico-chemical properties of DNAPLs and aquifer materials by (6-7): (1) reducing the interfacial tension between the wetting (generally the aqueous) phase and the non-wetting DNAPL phase; (2) reducing the viscosity of the DNAPL thereby promoting favorable mobility ratios for increased mobilization; or (3) enhancing solubilization of the DNAPL into surfactant micelles. The first two mechanisms can mobilize the DNAPL by releasing trapped oil from residual saturation or by causing the DNAPL and ground water to form a middle phase microemulsion (microemulsification). The third mechanism (enhanced solubilization) can result in DNAPL solubilities several orders of magnitude greater than the normal aqueous phase solubility of the DNAPL.

Surfactant-enhanced remediation processes will require several basic hydraulic steps including: (1) introduction of surfactant solutions to the subsurface; (2) effecting intimate contact between the surfactant solution and the contaminant; and (3) extraction of the resulting surfactant-contaminant mixture. During each of these steps, surfactants may be subject to losses due to physical and chemical reactions of the surfactant with subsurface materials. Significant masses of surfactant may be precipitated or sorbed in the subsurface (7-9). Also of concern in a surfactant-aided aquifer restoration program is the potential loss of surfactants to uncontaminated portions of the aquifer and the associated chemical costs (10).

Hydraulic Control Measures

The technical and economic feasibility of any surfactant-based remediation process will depend on the ability to achieve *hydraulic control* over the subsurface. Concurrently, it will be necessary to achieve hydraulic control while maximizing *hydraulic efficiency*. Hydraulic efficiency can be increased by: (1) minimizing the volume of injected surfactant solution; (2) minimizing the volume of fluid to be pumped to the surface (reducing treatment costs); (3) targeting injected chemicals to the contaminated zones of the aquifer; (4) preventing the movement of injected fluids towards clean portions of the aquifer; and (5) maximizing capture of resulting water-surfactant-contaminant mixtures.

The hydraulic efficiency of surfactant-aided injection/extraction can be dramatically increased by strategically locating and/or operating *impermeable* and/or *hydraulic* barriers. Impermeable physical barriers (e.g. grout curtains, slurry walls, sheetpiling) can be used to deflect flows into or away from contaminated zones by creating zones of low permeability. Hydraulic barriers (e.g. injection wells, infiltration galleries) can be used to deflect flows into or away from contaminated zones by creating zones of increased hydraulic potential (head). Several authors have also proposed the use of certain operational measures such as cyclic (pulsed) pumping, push-pull pumping, and variable injection/extraction ratios to improve pump-and-treat efficiency (*4*).

A recent numerical modeling study assessed the relative effectiveness of hydraulic and impermeable barriers for improving the efficiency of DNAPL remediation processes, both with and without surfactants (*11*). Simple injection of water can improve DNAPL extraction efficiencies but is hydraulically inefficient. Impermeable barriers accompanying injection and extraction wells dramatically improve DNAPL extraction efficiency by increasing the gradient through the contaminated zone and by reducing the volume of fresh ground water reaching the extraction wells. The overall conclusion drawn from these results was that mass transfer of the contaminant from the sorbed phase to the "*fluid*" moving through the contaminated zone should be maximized, regardless of whether the fluid is air, water, or a chemical solution.

Using a surfactant solution as the fluid offers the potential for dramatically improved mass transfer processes. Simple upgradient injection of surfactants followed by downgradient extraction is tremendously inefficient because a significant mass of surfactant is lost to uncontaminated zones and/or does not move through the contaminated zone. Injection of surfactant solutions inside partially encircling impermeable barriers with downgradient deflector wells was found to be most efficient for the surfactant-based processes. The impermeable barrier "cuts off" upgradient water (eliminates dilution of surfactant solution) and prevents migration of surfactant solutions into uncontaminated areas. The hydraulic barriers (deflector wells) provide increased gradient in addition to directional control. The volumes (mass) of surfactant solution required to exceed the critical micelle concentration (CMC) in the contaminated zone decreased significantly (up to 65%) with barriers over simple injection/extraction (*11*).

Pulsed pumping was first proposed by petroleum engineers to improve recovery from hydrocarbon reservoirs (*12*). However, since pulsed pumping has a resting phase, it may not increase the overall mass removal efficiency in remediation applications in terms of time. Disadvantages associated with pulsed pumping that have been identified in laboratory and field studies include increased remediation times, operation and maintenance issues, and lack of necessary hydraulic control (*13-15*).

Vertical Circulation Wells

Simultaneous injection to and extraction from a common vertical borehole creates a circulating flow pattern (Figure 1) within a sphere or ellipsoid around the borehole. These systems are referred to as vertical circulation wells (VCWs). The potential benefits of the VCW system are many and varied. The VCW system could be applied to DNAPL contamination by injecting a surfactant solution through one screened interval and extracting the surfactant/contaminant mixture from the other screened interval. Some benefits of the VCW system include: (1) reduced costs over

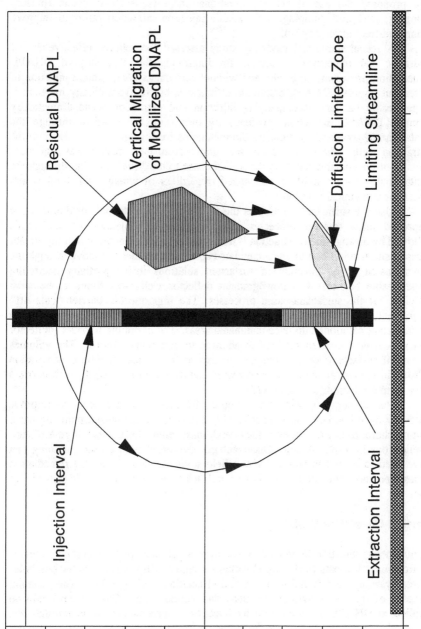

Figure 1. Vertical circulation well flow pattern.

systems involving multiple wells; (2) effective hydraulic control is achieved over limited volumes of the formation; (3) ability to capture NAPL's that might sink when mobilized; (4) can apply the system to both Light NAPLs (LNAPL's or "floaters") and DNAPL's ("sinkers"); (5) minimal loss of surfactants; (6) reduced volumes of fluids requiring treatment produced at surface; and (7) induced mounding can remediate portions of the contaminated vadose zone around the well.

Steady state flow induced by the VCW system in an aquifer with a regional gradient can be described using the complex potential, Ω,

$$\Omega = \Phi + i\, \Psi \tag{1}$$

where, Ω is the complex potential, Φ is the hydraulic potential, and Ψ is the stream function. Lines of constant Φ are called equipotentials and they describe the head distribution within the aquifer. Lines of constant Ψ are called streamlines and they describe the flow paths of ground water within the aquifer.

Referring to Figure 2a, the two screened intervals behave as a line source and a line sink, respectively. By superposition the complex potential for the line source and line sink can be combined, along with the complex potential for a regional gradient (horizontal flow), to produce the overall complex potential for a vertical slice of the aquifer. Using the equations for line sources/sinks and regional gradient developed by Strack (*16*), the complex potential becomes:

$$\Omega = \frac{\sigma D}{4\pi}[(Z - \frac{a}{D})\ln(Z - \frac{a}{D}) + (Z + \frac{a}{D})\ln(Z + \frac{a}{D})$$

$$- (Z - 1)\ln(Z - 1) - (Z + 1)\ln(Z + 1)] \tag{2}$$

$$- Q_o\, Z\, (\frac{D - (z_4 - z_1)}{2})$$

where, $\quad z = x + iy$
$\quad\quad\quad Z = X + iY$
$\quad\quad\quad Q_o$ = horizontal Darcy velocity
$\quad\quad\quad \sigma$ = strength of injection/extraction interval

and

$$Z = \frac{[z - \frac{1}{2}(z_4 + z_1)]}{\frac{1}{2}(z_4 - z_1)} \tag{3}$$

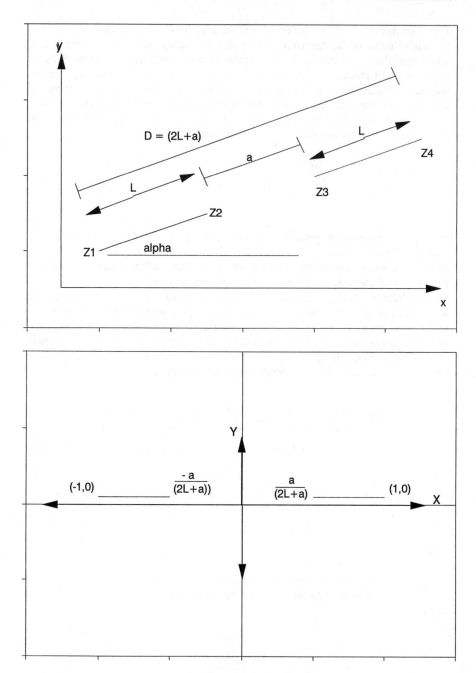

Figure 2. VCW system geometry: (a) VCW geometry in complex (z) plane (standard x-y coordinates); (b) transformed VCW geometry in dimensionless complex (Z) plane (dimensionless X-Y coordinates).

The variable transformation from z to Z is simply a change from the global (x,y) coordinate system to a local (X,Y) system based on the geometry of the vertical circulation well (see Figure 2a and 2b).

Figure 3 is a plot of the streamlines in an aquifer with a local regional gradient and a VCW system in which the bottom screened interval is extracting at a rate higher than the upper screened interval is injecting ($Q_{out} > > Q_{in}$). This allows for complete capture of the injected solution; however, the solution is significantly diluted by fresh ground water. This will result in diluted apparent effluent concentrations for both the injected surfactant solution and the solubilized (or mobilized) contaminant. It is important to recognize that the performance of the surfactant solution relative to extraction of the contaminant is masked in the VCW system effluent by dilution due to the fresh ground water that is extracted. It is important to note that this analysis does not consider the effects of surfactant solutions or NAPL on the conductivity of the media; rather, it simply describes the flow regime for a VCW in a homogeneous aquifer.

Physical Modeling Studies

The relative performance of the VCW system versus the traditional injection/extraction (two well) system (Figure 4) was assessed using a two-dimensional (sand tank) model packed with glass beads. The tank is constructed of aluminum with a glass front plate. The tank is 36 inches wide by 18 inches high, with a 2 inch interval separating the back panel and the from glass plate. The tank has variable constant-head end reservoirs, a glass front plate for visual observation, and piezometers for sampling and head measurement. In all simulations, a known mass of DNAPL was spilled into the tank and removal was attempted using surfactant solutions. DNAPL concentrations in the extracted effluent were monitored versus the volume of surfactant injected. Each system was assessed using both surfactant remediation mechanisms; enhanced solubilization and mobilization (microemulsification) using the same surfactant solutions. The VCW system was tested using two different modes of operation; circulating down and circulating up.

The surfactant solution used for the solubilization studies was 4.56 weight percent (wt%) solution of TMAZ 20, a non-ionic surfactant. The interfacial tension of this solution was estimated from previous studies to be about 5 dyne/cm. The solution used for the mobilization studies consisted of 1.3 wt% AOT and 3.6 wt% SMDNS, which produced an interfacial tension of 5×10^{-3} dyne/cm. AOT is an anionic surfactant and SMDNS is referred to as a "hydrotrope" and serves to increase the solubility of the AOT. The DNAPL used in both the mobilization and solubilization studies was PCE.

Two-Well System. Plotted in Figure 5 are the breakthrough and mass recovery curves for the two well system using mobilization and solubilization. The breakthrough curves show that mobilization results in higher effluent DNAPL concentrations than solubilization (Figure 5a) and thus quicker extraction of the DNAPL mass (Figure 5b). However, vertical migration of the mobilized DNAPL to the bottom of the tank resulted in non-extractable mass. This is evidenced by the

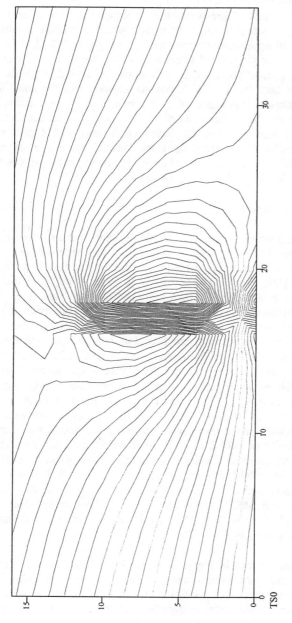

Figure 3. Equipotentials and streamlines for a VCW system in an aquifer with a regional gradient.

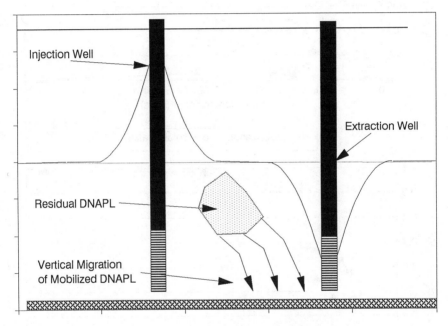

Figure 4. Two-well injection extraction system.

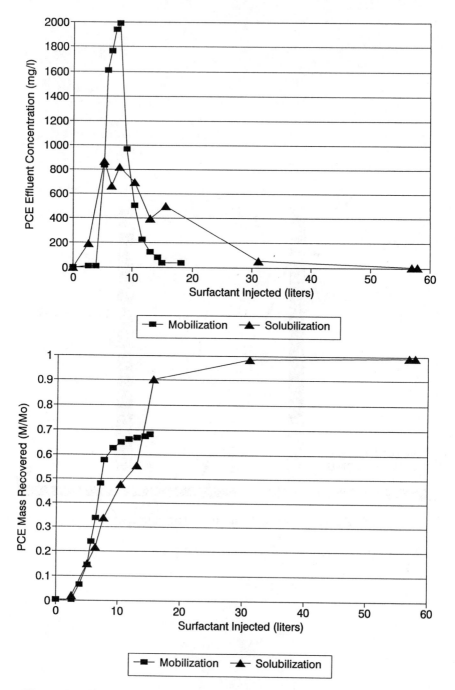

Figure 5. Effluent breakthrough and mass recovery curves for two-well system: (a) effluent breakthrough curve; (b) mass recovery curve.

mobilization mass recovery curve achieving only about 70% total mass recovery. In addition, the radial flow pattern emanating from the injection well in the two-well system results in significant surfactant mass lost to non-contaminated portions of the model.

Vertical Circulation Well System. Plotted in Figure 6 are the breakthrough and mass recovery curves for the VCW system using solubilization and mobilization. The breakthrough curves (Figure 6a) show that mobilization results in higher effluent DNAPL concentrations than solubilization. However, vertical migration of the mobilized DNAPL resulted in the accumulation of a diffusion-limited mass of contaminant at the fresh water-surfactant solution interface (i.e., the outermost streamline). Because the fluid outside the streamline (i.e., the ground water) contains no surfactants, the surfactant system accumulating at the interface is no longer a middle phase microemulsion. Slow dissolution of this diffusion-limited mass of contaminant negates the advantages of the higher concentrations achieved by mobilization. This is reflected in the long tailing of the mobilization mass recovery curve (Figure 6b) after an initial dramatic increase. It should be noted that if this phenomenon had not been observed, the mass recovery during mobilization would have likely been achieved within 3 to 4 liters versus 30 liters for solubilization. It is suggested that modifications to the operating conditions (e.g., flow reversal) can overcome this phenomenon.

Solubilization. Plotted in Figure 7 are the breakthrough and mass recovery curves for the two-well and VCW systems using the surfactant-enhanced solubilization mechanism (i.e., the same results as above but now plotting solubilization via two well and VCW jointly). The breakthrough curves (Figure 7a) show that the VCW system achieves higher effluent concentrations due primarily to reduced dilution by uncontaminated ground water in the effluent. The circulating flow pattern of the VCW also ensures higher recovery of the injected surfactant. However, the tailing of the VCW mass removal curve (Figure 7b) negates the advantage of higher concentrations. Again, operational modifications can be used to overcome this phenomenon.

Mobilization. Plotted in Figure 8 are the breakthrough and mass recovery curves for the two-well and VCW systems using the surfactant-enhanced microemulsification mechanism. The breakthrough curves (Figure 8a) appear to show that the two-well system achieves higher effluent concentrations than the VCW system. This is probably due to re-distribution of the mobilized DNAPL, i.e., the mobilized contaminant that actually reaches the extraction well approaches the well as a vertical wall of contaminant. However, as shown in the mass recovery curve (Figure 8b), the two-well system loses some of the mobilized residual to vertical migration.

Conclusions and Recommendations

Compared to the two-well system, the VCW system has the distinct advantage of

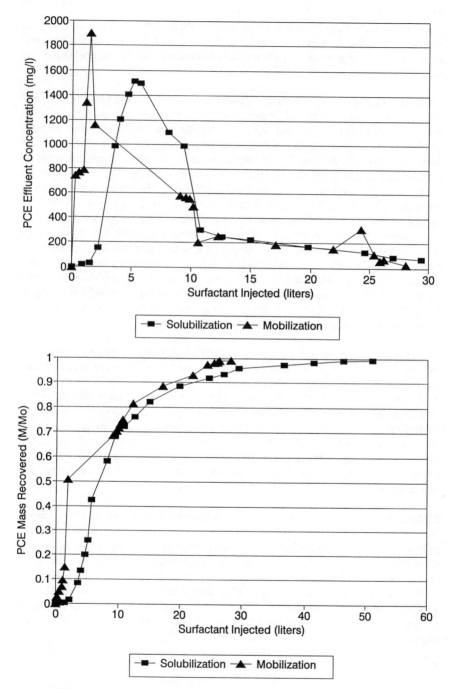

Figure 6. Effluent breakthrough and mass recovery curves for VCW system: (a) effluent breakthrough curve; (b) mass recovery curve.

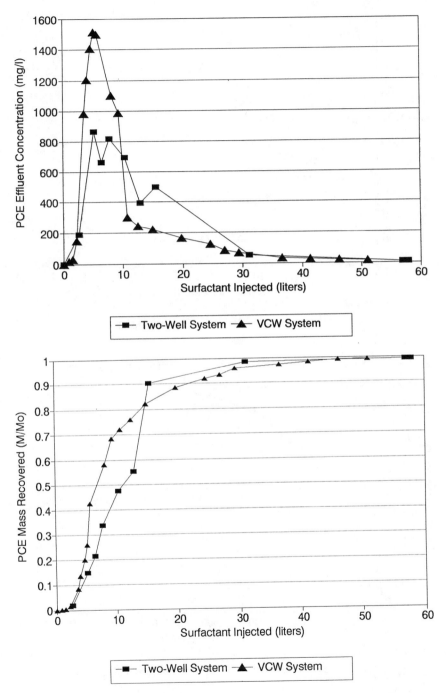

Figure 7. Effluent breakthrough and mass recovery curves for surfactant-enhanced solubilization: (a) effluent breakthrough curve; (b) mass recovery curve.

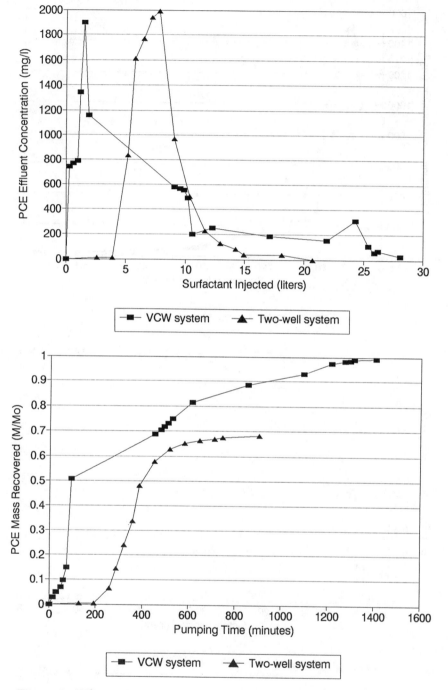

Figure 8. Effluent breakthrough and mass recovery curves for surfactant-enhanced microemulsification: (a) effluent breakthrough curve; (b) mass recovery curve.

hydraulic control over the mobilized residual. However, the VCW system **must** capture all of the mobilized residual in order to maintain this advantage (i.e., avoid loss of middle phase microemulsion at outer streamline). It is suggested that this can be accomplished by operational modifications such as flow reversal. These modifications will be evaluated in future research. In addition, without complete extraction of the mobilized residual, the solubilization mechanism is as hydraulically efficient as the mobilization mechanism using the VCW system. The VCW system also has the distinct advantage of higher surfactant recovery compared to the two well system. The middle phase microemulsion system has the potential to be significantly more efficient (i.e., an order of magnitude) when these operational problems are addressed.

The physical (*visual*) model is an effective means of conveying concepts and developing gross quantitative relationships. The mobilization mechanism can cause vertical migration of DNAPL; this must be understood and designed for (avoided or design the hydraulic regime to capture). Surfactant use, losses, and recovery will influence economic viability of all processes. Further evaluation of hydraulic control capabilities are needed in fully 3-dimensional flow systems. Other innovative measures (operational controls, extraction systems) need to be evaluated. Finally, simple mathematical (analytical) modeling such as that presented above allows for preliminary evaluation of the performance of the VCW system. However, numerical techniques will be required to evaluate field applications because the flow field is truly three-dimensional.

Literature Cited

1. U.S. Environmental Protection Agency, *Transport and Fate of Contaminants in the Subsurface*, EPA/625/4-89/019, Sept. 1989.

2. Miller, C.T., Poirier-McNeill, M.M., and Mayer, A.S., *Water Resources Research*, 1990, Vol. 26, No. 11, pp. 2783-2796.

3. Knox, R.C. and Sabatini, D.A., *Transport and Remediation of Subsurface Contaminants: Colloidal, Interfacial, and Surfactant Phenomena*, D.A. Sabatini and R.C. Knox, eds., ACS Symposium Series 491, American Chemical Society, Washington, DC, 1992, pp. 1-11.

4. Keeley, J.F., "Performance Evaluation of Pump-and-Treat Remediations", U.S. Environmental Protection Agency, EPA/540/4-89/005, Ada, OK, 1989.

5. Haley, J.L., Hanson, B., Enfield, C. and Glass, J., *Ground Water Monitoring Review*, Winter 1991, pp. 119-124

6. Mercer, J.W., and Cohen, R.M., *Journal of Contaminant Hydrology*, Vol. 6, 1990, pp. 107-163.

7. Harwell, J.H., *Transport and Remediation of Subsurface Contaminants: Colloidal, Interfacial and Surfactant Phenomena*. Sabatini, D. A. and Knox, R. C., eds., ACS Symposium Series 491, American Chemical Society, Washington, DC, 1992, pp. 124-132.

8. Palmer, C.D. and Fish, W., "Chemical Enhancements to Pump-and-Treat Remediation", U.S. Environmental Protection Agency, EPA/540/S-92/001, Ada, OK, 1992.

9. Rouse, J. D., Sabatini, D. A. and Harwell, J. H., *E S & T* In press (Accepted March 1, 1992).

10. Harwell, J.H. and Krebs-Yuill, B., *Proceedings of the Third International Conference on Ground Water Quality Research*, National Center for Ground Water Research, Houston, TX, 1992, pp. 230-232.

11. Gupta, H.S.,"Modeling the Effectiveness of Barriers for Improving the Hydraulic Efficiency of a Pump and Treat System Utilizing Chemical Agents."Master's thesis, University of Oklahoma, Norman, OK, 1993.

12. Aguilera, R., *Naturally Fractured Reservoirs*, Petroleum Publishing Company, Tulsa, OK, 1980.

13. Stallard, M. and Anderson, E., *Aquifer Restoration: Pump-and-Treat and the Alternatives*, AGWSE Educational Program, Las Vegas, NV, 1992.

14. Armstrong, J.E., Frind, E.O. and McClellan, R.D., *Water Resources Research*, 1994, Vol. 30, No. 2, pp. 355-368.

15. Voudrias, E.A. and Yeh, M., *Ground Water*, 1994, Vol. 32, No. 2, pp. 305-311.

16. Strack, O.D.L., *Groundwater Mechanics*, Prentice Hall, 1989.

RECEIVED January 23, 1995

Chapter 19

Economic Considerations in Surfactant-Enhanced Pump-and-Treat Remediation

B. Krebs-Yuill[1], J. H. Harwell[1,3], D. A. Sabatini[2], and R. C. Knox[2]

[1]School of Chemical Engineering and Materials Science and [2]School of Civil Engineering and Environmental Science, University of Oklahoma, Norman, OK 73019

Surfactants can aid pump-and-treat remediation of DNAPL chemicals but, under the conditions examined in this paper, surfactant re-use is necessary to be economical. The most cost effective above ground processes for surfactant regeneration are probably vacuum steam stripping or air stripping/incineration with surfactant recovery from the bleed stream using a combination of ultrafiltration and foam fractionation. The major cost is that of surfactant required to fill the treated zone. If the region of residual saturation is small and well defined, then surfactants may be more economical than pump-and-treat alone.

There is a growing interest in the use of surfactants to enhance the remediation of contaminated aquifers. Pump-and-treat technology is commonly employed to remediate contaminated groundwater formations. The main problem with this technology is the long time that may be required to reach acceptable cleanup levels. This is particularly true for dense non-aqueous phase liquid (DNAPL) contamination where a residual portion of the organic phase has become trapped in pore spaces of the formation due to capillary forces. The amount of organic contaminant which may be retained in the soil has been reported to range from 10% to 50% of the total pore space. (1) The objective of this study is to examine surfactant-enhanced pump-and-treat remediation and propose guidelines for its economic use in DNAPL remediation.

Using simple pump-and-treat remediation, the removal rate of residual DNAPL is limited by the rate and extent to which the DNAPL can dissolve. In addition, some water never contacts the residual organic and this further reduces the concentration in the recovered water. Because of these factors, recovered water usually contains DNAPL concentrations at about 10 percent of their aqueous solubilities. Residual tetrachloroethylene (PCE), with an aqueaous solubility of 200 mg/L, can require hundreds of years to remediate using conventional pump-and-treat methods. (1,2)

[3]Corresponding author

0097–6156/95/0594–0265$12.00/0

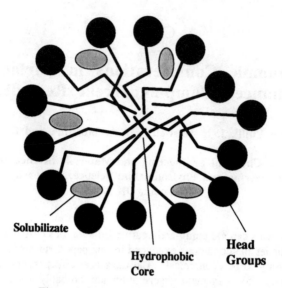

Figure 1. Representation of a micelle.

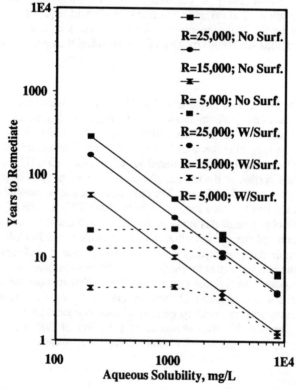

Figure 2. The remediation time required for very insoluble contaminants can be significantly reduced by using 1.7 wt. percent surfactant. R = ratio of mass of contaminant in aquifer to water removal rate (lb/gpm).

Surfactants have the potential to increase the aqueous solubility of DNAPLs. As shown in Figure 1, above the surfactant's critical micelle concentration (CMC), micelles form; they can be pictured as "oil" droplets with permeable hydrophillic shells and DNAPL dissolved in their cores. As the concentration of surfactant increases, the amount of DNAPL that can be solubilized increases because the number of micelles increases. Increased DNAPL solubility will reduce remediation times by increasing the rate of DNAPL removal for the same volume of water pumped. A well engineered surfactant-enhanced pump-and-treat process can offer efficient and low cost remediation of residual DNAPL.

Applicability of Surfactant Injection

Recovered water from a pump-and-treat operation, where a residual concentration of DNAPL exists, generally contains DNAPL concentrations well below the aqueous solubility (e.g., 10% of the aqueous solubility). If the extracted DNAPL concentration and the pumping rate are both assumed constant, then the time required to remediate a given mass of residual DNAPL would be proportional to its aqueous solubility. This is plotted in Figure 2 for various R values, where R is the ratio of the total mass of the particular DNAPL in the soil to the recovery water flowrate (pounds per gallon per minute). Aqueous concentrations are assumed to be 10% of their solubility limit. Also shown in Figure 2 is the remediation time required if surfactants are injected. It can be seen that surfactants have a pronounced effect for DNAPL species having an aqueous solubility of less than about 3000 mg/L.

The surfactant curves in Figure 2 were generated assuming that the recovered water has a sodium lauryl sulfate surfactant concentration of 1.7 wt. percent or 10 times its CMC (CMC = 0.006 M) (*3*) and the concentration of DNAPL in the surfactant micelles is in equilibium with an aqueous solution containing DNAPL at ten percent of its aqueous solubility. The Molar Solubility Ratio (MSR) is the molar ratio of DNAPL to surfactant in the micelles at aqueous saturation. An example determination of an MSR value is shown in Figure 3.

Given an MSR value, an equilibrium partitioning coefficient can be calculated and used to estimate the concentration of DNAPL in the micelles for any concentration of surfactant and aqueous phase DNAPL. The MSR values for this analysis were assumed to be 0.275 for tetrachloroethylene (PCE, aq. solubility = 200 mg/L), 0.20 for trichloroethylene (TCE, aq. solubility = 1100 mg/L) (*4*), 0.15 for 1,1,2,2-tetrachloroethane (aq. solubility = 2900 mg/L), and 0.10 for 1,2-dichloroethane (1,2-DCA, aq. solubility = 8690 mg/L). In actual applications the MSR will depend on temperature and electrolyte concentration (*5*). The relationships shown in Figure 2 should be reasonable estimates for sodium lauryl sulfate and other surfactant systems.

Another major concern with surfactant-enhanced pump-and-treat technology is the initial cost of surfactant required to fill the aquifer. Figure 4 shows this initial surfactant cost for PCE remediation of varying aquifer volumes at two different surfactant prices and two R values (ratio of PCE in the aquifer to water removal rate). The surfactant concentration used is set to theoretically allow complete remediation of PCE in seven years. An aquifer volume of 50 million gallons corresponds roughly to 25 acres by 20 feet at 30 percent porosity; 10 million gallons corresponds to 5

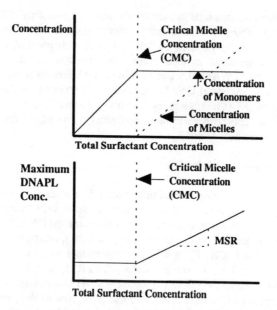

Figure 3. An example determination of the Molar Solubility Ratio (MSR).

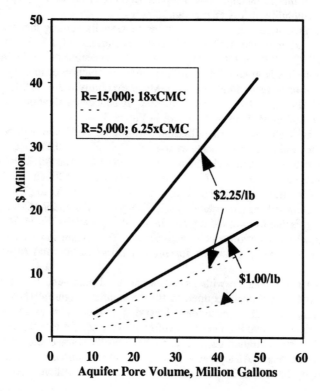

Figure 4. Initial surfactant cost as a function of aquifer pore volume.

acres with the same thickness and porosity. The surfactant costs in Figure 4 reflect only filling the void with surfactant-containing solution and do not consider surfactant losses due to sorption or precipitation. This graph emphasizes the value of defining the area containing residual DNAPL and confining the surfactant flood to that area.

It should also be emphasized that surfactants or surfactant mixtures should be chosen and tested for the particular remediation site. The surfactant chosen should exhibit minimum soil adsorption and precipitation.

Process Options

For residual saturations of low-solubility DNAPLs, the most viable remediation technology may be a surfactant-enhanced pump-and-treat process. Process economics dictate that the above-ground processing of the recovered water must be compatible with surfactant recovery and must minimize the cost of using surfactant.

The above ground process must treat the recovered water to remove DNAPLs. Treated water can then be used for reinjection. Since economics require that the surfactant be reused, this step is critical to the feasibility of the process. Typically, recovery wells will produce more water than is injected to insure hydrolic control DNAPLs and surfactants. This can result in a significant bleed stream, depending on the hydrology. Economics also dictate that surfactant be recovered from the bleed stream, but not from the entire recovery stream if this water can be reinjected. Figure 5 demonstrates the cost of surfactant losses without a recovery step from the bleed stream. As will be discussed later, the cost of surfactant recovery is only a small fraction of the cost of lost surfactant.

In order to identify the most economic surfactant recovery process, a base case was studied. In the base case, DNAPL must be removed from 500 gpm of recovery well water, the bleed stream is 150 gpm, and 350 gpm of treated water is reinjected along with surfactant. It is assumed that DNAPL concentrations in the reinjected water should be at or below the 0.5 ppb level. The treatment process will be based on chemical/physical differences between the DNAPLs, surfactant and ground water.

As shown in Table I, the initial separation can be based on differences in micellar/aqueous properties (Option I), volatility (Option II), surfactant/solution properties (Option III), or organic/aqueous properties (Option IV). Potential unit operations to achieve each separation are listed in Table II. Option II minimizes the number of unit operations as well as the total volume of water processed. This general approach is recommended and is shown in Figure 6.

Option IV also minimizes the number of unit operations, but processes to accomplish DNAPL removal to the desired levels for reinjection could not be identified. An appropriate solvent extraction followed by selective distillation may be possible.

Two process schemes, based on Option II, were identified and evaluated. If the recovered DNAPL has no value, standard air stripping of the process feed will remove the DNAPL and leave the surfactant for reinjection. Catalytic incineration of the DNAPL is followed by caustic scrubbing to remove HCl. The surfactant is recovered from the bleed stream using a combination of ultrafiltration and foam fractionation. Ultrafiltration is most efficient down to the CMC level and foam fractionation is efficient below the CMC level.

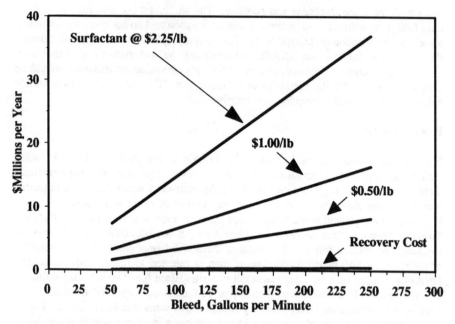

Figure 5. The cost of surfactant losses without a recovery step from the bleed stream.

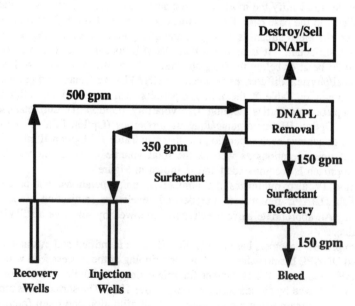

Figure 6. Recommended process approach for surfactant-enhanced pump-and-treat remediation.

Table I. General Approach Options

Stream	Option I	Option II	Option III	Option IV
Feed (500 gpm)	Separate Micelles	Separate All DNAPL	Separate All Surfactant	Separate All Organic
Concentrate (50 gpm)	Separate DNAPL from Surfactant			Separate DNAPL from Surfactant
Feed in Process(450 -500 gpm)	Separate Aqueous Phase DNAPL		Separate All DNAPL	
Bleed (150 gpm)	Recover Surfactant	Recover Surfactant		
DNAPL	Destroy or Sell	Destroy or Sell	Destroy or Sell	Destroy or Sell

Table II. Unit Operations Considered

Separate DNAPL From Feed
    Air Strip
    Steam Strip
    Vacuum Steam Strip
    Pervaporation
    Organic Extraction
    Biotreatment

Separate Surfactant From Bleed
    Waste
    Foam Fractionation
    Al,Ca Precipitation; Cation Exchange
    Micellar Enhanced Ultrafiltration (MEUF)
    MEUF combined with Foam Fractionation

Destroy or Sell DNAPL
    Incineration
    Carbon Adsorption
    Recycle/Sell/Store
    Biotreatment

If the DNAPLs have sufficient value to a reclaimer or can be recycled to an adjacent operation, vacuum-steam distillation becomes attractive. Surfactant is recovered from the bleed stream as in the previous case.

The initial design and cost estimates assume that surfactant will have little effect on the stripping efficiency. However, surfactant may change the partition coefficients, thereby decreasing the stripping efficiency. The air/water interface may also disrupt the micelles and actually increase the efficiency. Foaming may also result. Additional experimental work is needed in this area.

Cost Estimates

The hypothetical case studied represents a very large volume of contamination. It was assumed that 5 acres of contaminated soil, 20 feet in depth, has a porosity of 30 percent and the void fraction contains 13 volume percent DNAPL. The DNAPLs present were assumed to be equal volumes of PCE, TCE and 1,2-DCA. This equates to 5.74 million pounds of PCE, 5.16 million pounds of TCE and 4.46 million pounds of DCA. The surfactant was assumed to cost $1.00 per pound and was added in sufficient quantities to achieve 15 times its CMC in the recovered water. The recovery wells pump a total of 500 gpm and the bleed necessary was assumed to be 150 gpm. The recovery stream should contain 330 mg/L TCE, 400 mg/L PCE and 950 mg/L 1,2-DCA giving a required remediation time of seven years.

A total of 68 injection and recovery wells were costed at $10 per foot and $700 per well for pumps and miscellaneous. Piping to the treatment facility will be very site specific and was estimated quite conservatively. Estimated capital and operating costs for the wells and an air strip/incineration process are shown in Table III. When possible, vendors were contacted for equipment costs.

The DUALL Division of MET-PRO Corporation recommends three air strippers in series to treat 500 gpm of process water containing PCE, TCE and 1,2-DCA. The columns should be 9 feet in diameter with 28 feet of packing. Global Environmental estimates a catalytic incinerator operating at 1000° F with 30 ft^3 of Englehard catalyst will achieve 99 percent destruction efficiency. (Allied also manufactures a catalyst that can be used with high chloride streams.) The hot gas stream must be quenched to about 180° F. A 5-foot diameter caustic scrubber, with a design efficiency of 99.5 percent, is recommended for removing HCl from the cooled gas stream.

The cost of ultrafiltration and foam fractionation will depend on the specific surfactant used. Conservative costs are shown based on previous work at the University of Oklahoma. (6)

First year surfactant costs include the initial cost to fill the aquifer plus the surfactant that adsorbs to the soil. For the base case, adsorption is assumed to be 20 percent of the surfactant for the first two pore volumes. It is also assumed that the system is designed to avoid precipitation losses.

Table IV shows costs for the vacuum-steam stripping case. AWD, a division of DOW, estimates a ball park installed cost of $3 million for a 5-foot diameter column, package boiler (1500 to 2000 lb/hr steam), and instrumentation. These costs will increase rapidly if the water requires significant pretreatment for inorganics.

The base case costs assumed DNAPL concentrations resulting from a surfactant concentration in the treatment process feed of 15 times the CMC. However, the stripping equipment is sized mainly on flowrate and final DNAPL concentrations required. Therefore, an existing remediation project could potentially switch to surfactant addition and still utilize their DNAPL separation equipment. Additional DNAPL separation equipment (due to decreased efficiency with surfactant present), surfactant and surfactant recovery steps represent the only significant cost increases over what would be required without surfactant addition.

Table III. Cost Estimate for Air Strip/Incineration Technology With Surfactant Addition ($1,000's)

Capital

Injection/Recovery Wells		570
Air Strip and Scrubber		900
Incinerator, Catalyst		350
MEUF		500
Foam Fractionation		500
Instrumentation		250
Contingency		500
	TOTAL	$3,570

First Year Surfactant Costs

($1.00/lb)	$4,240

Annual Operating Costs

Well Operation		40
Fuel		45
Electricity		80
Maintenance (5% of capital)		180
Catalyst Replacement		20
Membrane Replacement		50
NaOH ($250/t)		400
Surfactant (10% of CMC from bleed)		125
Labor		100
	TOTAL	$ 940

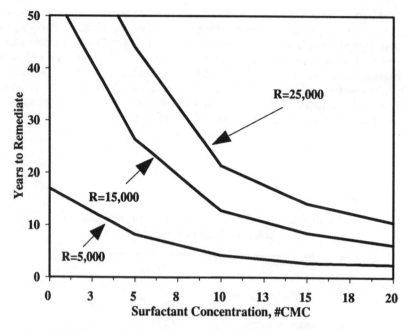

Figure 7. PCE remediation time as a function of surfactant concentration.

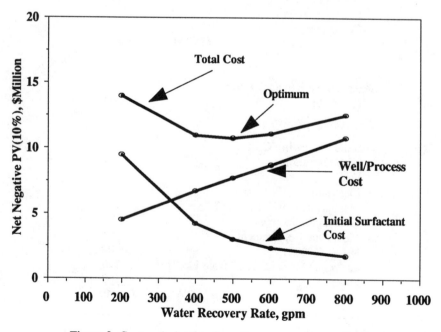

Figure 8. Cost optimization for a five acre surfactant flood.

**Table IV. Cost Estimate for Vacuum Steam Strip
Technology With Surfactant Addition ($1,000's)**

Capital		
Injection/Recovery Wells		570
Vacuum Steam Stripper		3,000
MEUF		500
Foam Fractionation		500
Contingency		500
	TOTAL	$5,070

First Year Surfactant Losses	
($1.00/lb)	$4,240

Annual Operating Costs		
Well Operation		40
Steam		120
Electricity		80
Maintenance (5% of capital)		250
Membrane Replacement		50
Surfactant (10% of CMC from bleed)		125
Labor		100
	TOTAL	$ 765

Project Optimization

The time required for remediation depends on the mass of DNAPL contamination, recovery well water flowrate, DNAPL aqueous solubility and the concentration of surfactant used. Figure 7 shows the time required for PCE remediation versus surfactant concentration for different R values. If the mass of PCE contamination is set, the three curves shown can be viewed as different ground water pumping rates. Lower pumping rates (higher R value) will require higher concentrations of surfactant to achieve remediation in the same time as higher pumping rates at lower surfactant concentrations.

For a given contamination site, this implies there is a cost tradeoff between the number of wells (assuming the optimum pumping rate per well is known and fixed), the time required for remediation, the size of the treatment process, and the initial surfactant cost. This is shown for the base case in Figure 8 where the above ground process is an air strip/incineration with surfactant recovery from the bleed stream. The negative net present value, discounted at ten percent, is plotted against the gpm of well water recovered. Surfactant concentration is varied to give a seven year remediation time; therefore, the initial surfactant cost decreases with gpm. The cost of wells and the above ground processing increases with increasing flowrate. For the base case contamination, the optimum flowrate, corresponding to the minimum negative net present value, is about 500 gpm. Initial surfactant costs due to adsorption losses were not included since they were assumed not to vary.

Sensitivity Studies

Since the major cost associated with surfactant-enhanced pump-and-treat remediation is the cost of the surfactant to fill the treated zone, it is important to define the area of residual saturation before implementing remediation efforts. If this area is small enough, the use of surfactants can be cost effective compared to pump-and-treat alone.

To illustrate this, consider a base case residual contamination of 198,000 lbs TCE, 222,000 lbs PCE and 171,000 lbs DCA with a 10 acre plume. This corresponds to equal volume portions of residual contaminant contained in 10 volume percent of 1/4 acre by 20 ft. For this Base Case example, pump-and-treat methods require 20 wells pumping at 5 gpm per well. The recovered water is treated using air stripping and incineration. Total remediation time is estimated to be 25 years and the project has a net negative present value, discounted at ten percent, of $1.7 million.

The sensitivity studies compare costs and remediation times required for this Base Case when using surfactant. The study variables are summarized in Table V. Case II assumes that residual DNAPL is evenly distributed over the 10 acres, thus the entire 10 acres requires application of surfactant. In Case III, the residual saturation is confined to only one acre and vertical circulation wells are used to apply surfactant and recover water from this one acre. Cases IV and V further confine the residual saturation and surfactant application to 1/2 and 1/4 acre, respectively. The plume is always recovered and treated.

Table V. Scope of Sensitivity Study

	Residual Coverage	Surfactant Applied to	Type Wells Used
Base Case	10 Acres	None	Standard
Case II	10 Acres	10 Acres	Standard
Case III	1 Acre	1 Acre	Vertical Circulation
Case IV	1/2 Acre	1/2 Acre	Vertical Circulation
Case V	1/4 Acre	1/4 Acre	Vertical Circulation

In this example, when residual saturation and surfactant application are confined to less than one acre, the net negative present value of the remediation project becomes equal to or less than the Base Case pump-and-treat. The estimated remediation times and economic comparisons are summarized in Table VI. The sensitivity study varied initial surfactant costs by changing the aquifer volume requiring surfactant. Other methods which reduce the initial surfactant costs should be equally effective for improving the economics of surfactant use. For example, methods which would reduce dilution of the surfactant solution (obtaining a closer approach to equilibrium solubilization) would allow injection of lower concentrations of surfactants. Development of the technology necessary to use mobilization with dilute (1 wt%), low pore volume slugs (10% of a pore volume) would also improve the economics.

Table VI. Economic Results of Sensitivity Study

		Years Required for Remediation	Negative NPV(10%) $1000
Base Case	No Surfactant	25	1,700
Case II	Surfactant to 10 Acres	2	8,700
Case III	Surfactant to 1 Acre	2	1,900
Case IV	Surfactant to 1/2 Acre	2	1,500
Case V	Surfactant to 1/4 Acre	4	1,400

End of Project Surfactant Recovery Option

Part of the initial surfactant cost can be recovered at the end of the remediation project. This would be accomplished by continued water flooding at a reduced rate and without surfactant addition. Only the surfactant recovery portion of the above ground process would be operated. The surfactant should retain a fair fraction of its original value, especially to another surfactant-enhanced remediation project. This may also be necessary if there is any environmental concern regarding leaving surfactant in the aquifer.

Conclusions

Surfactant-enhanced pump-and-treat remediation is effective for DNAPL chemicals with aqueous solubilities of less than 3,000 mg/L, but requires surfactant re-use to be economical. The most cost effective above ground processes for surfactant regeneration are vacuum steam stripping or air stripping/incineration, with surfactant recovery from the bleed stream. This is based on the assumption that surfactant will have little effect on the stripping efficiency. The best surfactant recovery process is probably a combination of microfiltration and foam fractionation.

If an arbitrary clean-up deadline is set, use of surfactants in a pump-and-treat project may be more economical depending on the site size, residual saturation, hydrophobicity of the contaminant, and the cost and concentration of the surfactant. Also, if the region of residual saturation is small and well defined and if pump-and-treat alone would require more than 15 to 20 years, then surfactants may improve economics of pump-and-treat.

The big difference between pump-and-treat and surfactant-enhanced pump-and-treat is the cost of the surfactant to fill the treated zone. Methods of reducing dilution of the surfactant solution will improve the economics by allowing use of lower concentrations of surfactants. Development of the technology necessary to use mobilization with dilute, low pore volume slugs will also improve the economics.

Literature Cited

1. *Dense Nonaqueous Phase Liquids - A Workshop Summary*; EPA/600/R-92/030; U.S. Environmental Protection Agency: Ada, Oklahoma, February 1992.
2. Mercer, J. W.; Skipp, D. C.; Giffin, D.; *Basics of Pump-and-Treat Ground-Water Remediation Technology*; EPA-600/8-90/003; U.S. Environmental Protection Agency: Ada, Oklahoma.
3. Rosen, M.J.; *Surfactants and Interfacial Phenomena*; Second Edition; John Wiley and Sons; 1989.
4. Sabatini, D. A.; R. C. Knox; *Transport and Remediation of Subsurface Contaminants*; ACS Symposium Series 491; 1992.
5. West, C. C.; Harwell, J. H.; *Environ. Sci. Tech.* **1992**, *26(12)*, p. 2324.
6. Dunn, R. O. Jr.; Scamehorn, J. F.; Christian, S. D.; *Sep. Sci. Tech.* **1985**, *20(4)*, pp. 257-284.

RECEIVED November 29, 1994

PANEL DISCUSSION

Chapter 20

Surfactant-Enhanced Remediation of Subsurface Contamination

Review of Emerging Technologies and Panel Recommendations

Candida C. West

Robert S. Kerr Environmental Research Laboratory, U.S. Environmental Protection Agency, Ada, OK 74820

These are exciting and extremely challenging times for those of us who are professionals in the field of subsurface remediation. We have been charged to bring to practice innovative technologies that have shown substantial promise for improving the way we currently practice reclamation of ground water to regulatory acceptability. Much of the current pressure for the development of innovative technologies stems from the widespread recognition that, with the exception of the case of dissolved plumes in homogeneous media, we are largely unable to clean up sites to regulatory contaminant levels using current technology. Many innovative technologies are currently at various stages of development including those that use the addition of chemical amendments to the extraction fluid for the purpose of chemically altering the way contaminants partition from aquifer solids and pore spaces into the mobilizing fluid. Surfactants are one class of chemical additives that may be used successfully to enhance the remediation process, particularly pump-and-treat technology as it is currently practiced. Surfactant enhanced subsurface remediation has been identified as one of the technologies worthy of serious evaluation. Over the last few years, the growth in the number of government, academic, and industry research laboratories conducting research on some aspect of surfactant-based remediation has been short of incredible. There is tremendous momentum behind this effort which has been primarily focussed on developing the solid science base that is going to be required to bring these technologies to fruition. The energy of that momentum was evident by the number and quality of the presentations made at the two-day session of the "207th American Chemical Society Meeting" in San Diego.

Our Goal - A Public Mandate

The current list of contaminated sites that have been successfully remediated is woefully short. Those sites that have been successfully remediated are generally those where the contaminants are present in dissolved form and the geology is relatively homogeneous. Unfortunately, this is generally not the case at the majority of sites; and the complexity of the contaminant matrix and the geology have significantly hampered remedial action. In response to this realization, there is increased pressure by the public and the regulatory community to improve our performance in subsurface remediation in a timely and scientifically defensible manner. As part of the effort to address remedial needs, we are faced with the difficult task of recognizing and developing the best course of action to be taken to develop surfactant-based remediation into viable and valid remedial tools. Based on current needs for innovative remedial technologies, it appears that the greatest contribution surfactant-based remedial technologies can make is in the area of remediating nonaqueous phase liquids, particularly dense nonaqueous phase liquids for which there are currently no remedial tools available. To address this, it is necessary to identify the issues that must be resolved to develop this remedial tool in a logical and accomplishable order. To this end, an invited panel was assembled at the conclusion of the symposium for the purpose of providing a forum from which key issues crucial to the development of this technology could be identified and discussed. The panel represented several segments of the community having a commitment in surfactant-based remediation. The panel members were: Dr. Linda Abriola, Department of Environmental Engineering at the University of Michigan, and the principal investigator for several research projects in surfactant-enhanced aquifer remediation; Mr. James Greenshields of ICI Surfactants, a major manufacturer and distributor of surfactants; Dr. Abdul Abdul, a researcher with General Motors who has conducted laboratory and field research evaluating the use of surfactants for enhancing removal of contaminants; Dr. Jeffrey Harwell, Institute for Applied Surfactant Research, University of Oklahoma, who has an extensive background in surfactant use in industrial and environmental applications, including enhanced oil recovery; and the author, as a representative researcher for the U.S. Environmental Protection Agency. The audience was equally broad in its representation of these segments of the provider and user community.

What Course Do We Take? Panel Discussion

The Need for an Interdisciplinary Approach. It became clear through the symposium discussions that the most challenging aspect of surfactant-based technology development will be the necessity for creating a forum through which open collaboration and communication between experts from many disciplines can be achieved. There is a fundamental distinction between research that is conducted solely for the simple extension of fundamental knowledge and that which is conducted for the purpose of producing a finished technological product ready for market use. This effort will require the active participation of experts covering all aspects of the surfactant utilization process: microbiologists for the determination of surfactant biodegradability and specific metabolic pathways; toxicologists evaluation of the acceptability of injecting a chemical amendment into the subsurface environment given possible receptors; chemists and geologists for evaluation of the compatibility of surfactant solutions given specific water chemistry and geochemistry; hydrologists for designing injection/extraction systems that will provide adequate delivery of the surfactant to the contaminant and proper hydrologic recovery of the injection solution; and engineers for process development of appropriate recovery processes for surfactant reuse and contaminant treatment. The primary stakeholder in this business is the general public and it is our duty to work towards acceptable technologies that will aid in cleaning up our nation's ground water and not make a bad situation worse. We will be required to share ideas and results developed at the laboratory bench, at small-scale field demonstrations, and full-scale field projects.

Where Are We Now? When research in the area of surfactant-based remediation was begun, the logical question asked was whether or not using surfactants could enhance contaminant removal or destruction sufficiently to warrant further investigation. It would be accurate, I believe, to say that bench-scale laboratory experiments have shown that the enhanced removal of residual phase contaminants from ground water via solubilization or mobilization in surfactant solutions warrants further evaluation and development. If bench-scale experimental results were directly applicable to field situations, there could be realized one to several orders of magnitude reduction in the flush volume required to remove a given mass of contaminant using surfactant-amended pump-and-treat remediation. The potential to use surfactants to enhance bioremediation is perhaps less clear, due to the lack of a fundamental understanding of the mechanism by which surfactants cause microbial toxicity or aid in microbial utilization of contaminants.

The questions now posed are directed towards the development of fundamental data required to move this technology to small-scale field demonstrations. Can we adequately delineate the area of contamination to be remediated and deliver the surfactant solution to it effectively? If so, what percentage of the surfactant flush can we recapture? Given some escape rate, what will be the surfactant's fate and how far will it travel before it is reduced to safe concentrations given processes such as sorption and/or biodegradation? What is a "safe" concentration? What are the biodegradation products and what are their toxicity to the same set of receptors? Can the recovered surfactant be processed for reuse and can the contaminant be recovered from the surfactant solution for treatment? How might the surfactant affect other intrinsic or active remediation processes? These were the range of questions that were addressed by the panel and audience.

Analogies to EOR. There are those who would say that we should use the example set in enhanced oil recovery (EOR) as evidence that the use of surfactants to remediate aquifers should not be pursued. Surfactant flooding was developed by the oil industry as a way to improve tertiary recovery of oil deposits. Both the oil companies and the surfactant manufacturers put substantial research resources into the development of this technology which for various reasons never came to fruition. However, the motivations and the stakeholders are as different in these two applications as are the environments in which they are used. Enhanced oil recovery involved injection of massive volumes of surfactant solution into hostile environments of high temperature and pressure over areas typically orders of magnitude greater than proposed for ground-water remediation. Additionally, the economical feasibility was driven by the price of oil, which at its peak was approximately $50/barrel, whereas the recovery of a single barrel of a contaminant such as tetrachloroethylene may well run thousands to tens of thousands of dollars. Given these significant differences in application and return, the two scenarios are not directly comparable. It is, however, important to point out that environmental researchers have examined many of the theories developed for EOR application for the prediction of surfactant behavior and its usefulness in this field. An example of this is the relationship between the bond and capillary numbers for the prediction of the onset of mobilization, as conducted by Dr. Linda Abriola and her coworkers.

Surfactant Receptors, Fate, and Toxicity. There is legitimate concern over injecting surfactants into ground water and possibly trading one contaminant for another. To allay this concern, it will be necessary to develop an extensive database on the toxicity of surfactants and their breakdown products to possible receptors. Many surfactants are toxic to aquatic organisms at relatively low concentrations (on the order of low parts per million) and should perhaps not be considered in situations where the injection solutions could reach and impact any surface waters such as swamps, rivers, streams, or lakes. This will require calculated estimates of rates of loss through sorption and degradation, time of travel, and rates of dilution.

It is clear that surfactants will either need to be completely non-toxic to the receiving system or be known to biodegrade within a reasonable period of time. However, an acceptable rate of biodegradation will be completely dependent on the specific use of the surfactant. The surfactant solution must be stable for a sufficient period of time required to do the job for which it was intended, but not so recalcitrant as to represent a long-term contaminant in itself. Currently, most degradation data on surfactants are for aerobic systems. There are very little anaerobic data. If surfactants are to be considered for use as part of a treatment train concept, for instance as part of a biologically mediated process, the effect of the surfactant on the microbial population of the system needs to be determined.

It was pointed out that the need for an extensive database on biodegradation rates, products, and toxicity, combined with the need to have a thorough understanding of surfactant phase-behavior and chemical compatibility with various water and mineral geochemistries, may necessitate focussing on a narrow selection of surfactants that could be studied intensely and recommended for use. This would reduce the research costs associated with the development of this database. It seems clear, however, that regulatory guidelines for surfactant acceptability must be developed before the research community can focus on a select group of surfactants and essentially put all its eggs in a few baskets.

Nonaqueous Phase Liquids and Contaminant Delineation. Adequate contaminant delineation and characterization has been identified as crucial to successfully instituting surfactant-enhanced remediation. This is particularly true for dense, nonaqueous phase liquid (DNAPL) contamination for which surfactant remediation has been identified as a key innovative technology. DNAPLs are particularly difficult to detect either directly via soil coring due to their elusive nature (i.e. sinking deep into aquifers) or by inference based on dissolved aqueous-phase concentrations. Geophysical techniques for source delineation are being developed, but may be relatively expensive and may not be available in the near future. Another delineation technique using partitioning tracers appears to be a promising technique and may be more rapidly developed for use than geophysical techniques. Again, unless we can improve our ability to locate and delineate residual and free-phase liquids, surfactants cannot be properly delivered for the purpose of removal of the DNAPL.

Innovative surfactant-based remediation technologies cover a wide range of contaminant types, both chemical and physical. Many address the enhancement of the remediation of dissolved plumes either by increasing the bioavailability of the contaminant or immobilizing the contaminant making it available for subsequent abiotic or biotic *in situ* treatment. Remediation of dissolved plumes using modified or enhanced pump-and-treat remediation in relatively homogeneous media probably has a high likelihood of success, but represents a small fraction of the sites currently mandated for clean-up. One of the most serious and difficult situations for which there is no practical remedial solution, even under uniform, homogeneous aquifer conditions, is that of remediation of nonaqueous phase liquids (NAPLs). The limitation for remediation of NAPLs comes from the large volumes of material introduced into the environment relative to its aqueous solubility, posing the limiting factor for pump-and-treat remediation. The vast majority of Superfund and RCRA sites are contaminated with NAPLs, and it is projected that tens to thousands of years would be required to remove the volume of NAPL based on calculations of the mass removed per pore volume at contaminant saturation. The problem is exacerbated by the fact that saturation is rarely, if ever, achieved making the time required to remove these materials even greater. It is the remediation of NAPLs for which surfactants can make its greatest contribution, either through the process of solubilization or mobilization.

Field Demonstrations. As has been the case in the past, many researchers expressed the need for field sites which could qualify for use as research demonstrations for surfactant-based remediation such as the field site in Borden, Canada. It was recognized that the United States seems to be moving in this direction and that the DoD has provided funding for the development of several field research sites. It was recommended that researchers submit proposals which would provide collaborative, holistic small-scale demonstration approaches encompassing innovative techniques for site characterization, contaminant source delineation, injection/extraction well geometries, and treatment systems for surfactant reuse and contaminant removal and treatment. As part of this collaboration, a stepwise test protocol for small-scale field demonstrations needs to be developed and modified as field and laboratory data are collected and analyzed. These test sites would also provide an opportunity to study surfactant interactions at more complex sites than have been typically studied. Research intensive small-scale pilot demonstrations would then provide the information required to develop full-scale site remediation.

Closing Remarks

One of the major barriers to the development of innovative technologies is the risk of failure, or worse, the risk of exacerbating the situation. This risk barrier is evident by the user community's reluctance to approve installation and operation costs for an unproven technology while still being held liable for the installation of a "standard" technology if the innovative technology should fail. Nor do regulatory agencies feel compelled to share in the costs of supporting these new technologies. The panel discussion was closed with the suggestion to form an expert panel consisting of representatives of all of the disciplines and stakeholders previously mentioned for the purpose of evaluating and promoting surfactant-based remediation technologies. It would be the panel's responsibility to offer their expert services for evaluation of proposed surfactant-aided remediation projects. The availability of this service would help prevent irresponsible use of surfactant-based remediation technologies and promote communication between scientists, users, and the regulatory community. It was speculated that the formation of a cooperative research and development agreement (CRADA), spearheaded by a regulatory agency such as the EPA, might be an appropriate vehicle through which to form this panel. In this way, the panel could facilitate possible opportunities for both small-scale and scaled-up field tests which could provide the kind of scientifically defensible data required to demonstrate the economic feasibility and appropriate application of surfactant-based aquifer remediation.

The opportunities for research in and application of surfactant-based remediation are growing. In order for our efforts to be fruitful it will be necessary to continue communication through symposia such as this one and through the creation of expert panels as discussed previously. It was generally agreed upon by the panel discussion participants that there is a need to meet at least every other year and to continue to hold panel discussions for the purpose of interchange of ideas. The author would like to thank all of those who participated in the panel discussions. It is hoped this chapter is an accurate reflection of your ideas and suggestions and may be in some way useful to your endeavors in surfactant-enhanced remediation.

RECEIVED December 13, 1994

INDEXES

Author Index

Abriola, L. M., 10
Adeel, Zafar, 38
Allred, B., 216
Bai, Gui-Yun, 82
Barber, Larry B., II, 95
Bourbonais, Katherine A., 161
Bowman, Robert S., 54
Brown, G. O., 216
Brusseau, Mark L., 82
Butler, George W., 201
Byrne, Michael, 177
Chen, L. Y., 249
Chen, Y., 249
Christian, Sherril D., 231
Chu, Wei, 24
Compeau, Geoffrey C., 161
Dekker, T. J., 10
Field, Jennifer A., 95
Flynn, Matthew M., 54
Fountain, John C., 177,191
Frazier, Dave, 177
Freeze, G. A., 191
Haggerty, Grace M., 54
Harvey, Ron W., 95
Harwell, Jeffrey H., 1,65,124,265
Hoof, Patricia L. Van, 24
Huddleston, Roger G., 54
Jackson, Richard E., 191,201

Jafvert, Chad T., 24
Knox, Robert C., 1,249,265
Krebs-Yuill, B., 265
Krueger, Carolyn, 95
Lagowski, Alison, 177
Lueking, Donald R., 112
Luning-Prak, D. J., 10
Luthy, Richard G., 38
MacClellan, Lee K., 161
McNally, Dan L., 112
Metge, David W., 95
Mihelcic, James R., 112
Miller, Raina M., 82
Neel, Daphne, 54
Pennell, K. D., 10
Pickens, John F., 201
Pope, Gary A., 10,142,191,201
Rouse, Joseph D., 65,124
Sabatini, David A., 1,65,124,265
Scamehorn, John F., 231
Shiau, Bor-Jier, 65
Taylor, Craig, 177
Waddell-Sheets, Carol, 177
Wade, W. H., 142
Wang, Xiaojiang, 82
West, Candida C., 280
Yin, Yuefeng, 231
Zhang, Yimin, 82

Affiliation Index

AGI Technologies, 161
Carnegie Mellon University, 38
INTERA Inc., 191,201
Michigan Technological University, 112
National Oceanic and Atmospheric
    Administration, 24
New Mexico Institute of Mining and
    Technology, 54
Oklahoma State University, 216
Oregon State University, 95

Purdue University, 24
State University of New York—Buffalo,
    177,191
U.S. Environmental Protection Agency, 280
U.S. Geological Survey, 95
University of Arizona, 82
University of Michigan, 10
University of Oklahoma,
    1,65,124,231,249,265
University of Texas, 10,142,191,201

Subject Index

Production: Michelle D. Althuis & Charlotte McNaughton
Indexing: Deborah H. Steiner
Acquisition: Rhonda Bitterli
Cover design: Cornithia Allen Harris

Printed and bound by Maple Press, York, PA

Highlights from ACS Books

Good Laboratory Practice Standards: Applications for Field and Laboratory Studies
Edited by Willa Y. Garner, Maureen S. Barge, and James P. Ussary
ACS Professional Reference Book; 572 pp; clothbound ISBN 0–8412–2192–8

Silent Spring Revisited
Edited by Gino J. Marco, Robert M. Hollingworth, and William Durham
214 pp; clothbound ISBN 0–8412–0980–4; paperback ISBN 0–8412–0981–2

The Microkinetics of Heterogeneous Catalysis
By James A. Dumesic, Dale F. Rudd, Luis M. Aparicio, James E. Rekoske, and Andrés A. Treviño
ACS Professional Reference Book; 316 pp; clothbound ISBN 0–8412–2214–2

Helping Your Child Learn Science
By Nancy Paulu with Margery Martin; Illustrated by Margaret Scott
58 pp; paperback ISBN 0–8412–2626–1

Handbook of Chemical Property Estimation Methods
By Warren J. Lyman, William F. Reehl, and David H. Rosenblatt
960 pp; clothbound ISBN 0–8412–1761–0

Understanding Chemical Patents: A Guide for the Inventor
By John T. Maynard and Howard M. Peters
184 pp; clothbound ISBN 0–8412–1997–4; paperback ISBN 0–8412–1998–2

Spectroscopy of Polymers
By Jack L. Koenig
ACS Professional Reference Book; 328 pp;
clothbound ISBN 0–8412–1904–4; paperback ISBN 0–8412–1924–9

Harnessing Biotechnology for the 21st Century
Edited by Michael R. Ladisch and Arindam Bose
Conference Proceedings Series; 612 pp;
clothbound ISBN 0–8412–2477–3

From Caveman to Chemist: Circumstances and Achievements
By Hugh W. Salzberg
300 pp; clothbound ISBN 0–8412–1786–6; paperback ISBN 0–8412–1787–4

The Green Flame: Surviving Government Secrecy
By Andrew Dequasie
300 pp; clothbound ISBN 0–8412–1857–9

For further information and a free catalog of ACS books, contact:
American Chemical Society
Product Services Office
1155 16th Street, NW, Washington, DC 20036
Telephone 800–227–5558

Bestsellers from ACS Books

The ACS Style Guide: A Manual for Authors and Editors
Edited by Janet S. Dodd
264 pp; clothbound ISBN 0–8412–0917–0; paperback ISBN 0–8412–0943–X

Understanding Chemical Patents: A Guide for the Inventor
By John T. Maynard and Howard M. Peters
184 pp; clothbound ISBN 0–8412–1997–4; paperback ISBN 0–8412–1998–2

Chemical Activities (student and teacher editions)
By Christie L. Borgford and Lee R. Summerlin
330 pp; spiralbound ISBN 0–8412–1417–4; teacher ed. ISBN 0–8412–1416–6

Chemical Demonstrations: A Sourcebook for Teachers,
Volumes 1 and 2, Second Edition
Volume 1 by Lee R. Summerlin and James L. Ealy, Jr.;
Vol. 1, 198 pp; spiralbound ISBN 0–8412–1481–6;
Volume 2 by Lee R. Summerlin, Christie L. Borgford, and Julie B. Ealy
Vol. 2, 234 pp; spiralbound ISBN 0–8412–1535–9

Chemistry and Crime: From Sherlock Holmes to Today's Courtroom
Edited by Samuel M. Gerber
135 pp; clothbound ISBN 0–8412–0784–4; paperback ISBN 0–8412–0785–2

Writing the Laboratory Notebook
By Howard M. Kanare
145 pp; clothbound ISBN 0–8412–0906–5; paperback ISBN 0–8412–0933–2

Developing a Chemical Hygiene Plan
By Jay A. Young, Warren K. Kingsley, and George H. Wahl, Jr.
paperback ISBN 0–8412–1876–5

Introduction to Microwave Sample Preparation: Theory and Practice
Edited by H. M. Kingston and Lois B. Jassie
263 pp; clothbound ISBN 0–8412–1450–6

Principles of Environmental Sampling
Edited by Lawrence H. Keith
ACS Professional Reference Book; 458 pp;
clothbound ISBN 0–8412–1173–6; paperback ISBN 0–8412–1437–9

Biotechnology and Materials Science: Chemistry for the Future
Edited by Mary L. Good (Jacqueline K. Barton, Associate Editor)
135 pp; clothbound ISBN 0–8412–1472–7; paperback ISBN 0–8412–1473–5

For further information and a free catalog of ACS books, contact:
American Chemical Society
Product Services Office
1155 16th Street, NW, Washington, DC 20036
Telephone 800–227–5558